TEACHER EDUCATION in AGRICULTURE

Second Edition

TEACHER EDUCATION

in

a project of the

AMERICAN ASSOCIATION OF TEACHER EDUCATORS IN AGRICULTURE

Edited by **Arthur L. Berkey**

AGRICULTURE

The Interstate PRINTERS & PUBLISHERS, INC.
DANVILLE, ILLINOIS

Teacher Education in Agriculture

—

SECOND EDITION

Library of Congress Catalog Card No. 81-83265

ISBN 0-8134-2217-5

Preface

Numerous changes in the milieu of vocational education have occurred during the intervening period since the first edition of Teacher Education in Agriculture was published in 1967. These include changes in initiatives by the agricultural education profession, in legislation, in public sentiment, in the school age population, and in technology in the world of work.

These changes have implications for teacher education in agriculture in two areas: (1) new approaches, standards, and mandates intended to improve teacher education programs and (2) the changing role which the vocational agriculture teacher must be prepared to fulfill.

The competency/performance-based approach to teacher education is now used partly or totally in most institutions preparing agriculture teachers. Establishing quality standards for all phases of agricultural education, the Standards for Quality Programs in Agricultural/Agribusiness Education represents a significant step forward by the agricultural education profession. The national conference at Kansas City in 1980 on "Agricultural Education in the 1980's" provided direction for program planning in this decade.

At the same time, federal funding for vocational programs has fallen far behind inflation. Additionally, legislative mandates to provide services to students with disabling conditions and sex equity instruction to all students are now part of the teacher's role. Basic skills emphasis, energy conservation, and teaching methodology to meet the needs of students with disabilities who are mainstreamed into vocational classes also represent changes in teacher roles.

The movement toward teacher licensing and state qualification tests for teachers reflects public sentiment for increased accountability by public education at all levels.

The declining school age population has also altered the teacher's role--and necessarily teacher preparation for the new role. Smaller numbers of students mean more combined and flexible course content options. It also means that teachers of agriculture may be more involved in providing instruction to out-of-school groups than in the past. Similarly, the increasing technology in agriculture/agribusiness will predictably require teachers with higher levels of technical preparation.

The chapters of this revised edition of Teacher Education in Agriculture have been written by leaders in the profession. As a result, positions and priorities by the respective authors are varied. However, there is strength in this diversity, as reflected by the common emphasis in all chapters on high quality programs for teacher preparation in agriculture.

Thanks are due to the authors for their cooperation and endeavors which made possible this revised edition.

Arthur L. Berkey

Contents

CHAPTER 1

Development of Teacher Education in Agriculture*

W. Howard Martin
UNIVERSITY OF CONNECTICUT

Arthur L. Berkey
CORNELL UNIVERSITY

Agriculture is taught in about 9,000 secondary schools, 100 universities and colleges, and many junior colleges and technical institutions. In addition, informal out-of-school education in agriculture is made available to large numbers of persons through agents of the cooperative extension service in agriculture and home economics. College and secondary teachers of agriculture, as well as extension agents, may be regarded as agricultural educators.

Various standards control entry into agricultural education. Those engaged in teaching agriculture in secondary schools--and to some extent those in post-secondary vocational institutions--are required to have a teaching certificate. Also, in several states, for example, California, certification to teach at the two-year college level is required. Requirements for certification to teach agriculture include preparation in the discipline of education and preparation in the technical agriculture and liberal arts disciplines commonly included in curricula of students majoring in agriculture.

Teacher education in agriculture involves the cooperative efforts of university faculty members, cooperating teachers in public schools, and members of the state departments of education. The cooperative efforts are coordinated by teacher educators in agriculture at universities.

Focus on the development, organization, and administration of teacher education in agriculture is guided by a concern for the future as well as by an interest in the past. Although the precise nature of the future can only be dimly perceived, certain forces are already shaping its character. The problem is to recognize these emergent forces and to assess the direction and

---

*Dr. Martin is the author of the original chapter on development of teacher education in the 1967 edition. Revisions of the chapter in this edition are primarily the work of Arthur L. Berkey. The revisions consist primarily of updating the period of 1967-1981. The sections on development prior to 1967, which are a major part of the chapter, are essentially unchanged from the 1967 edition.

degree of their probable influence. Thus, attention is given to those forces of influence, past and emergent, as well as to the data regarding origins and patterns of organization and administration (1.1).

Origins and Establishment

When did teacher education in agriculture begin? Some say it began with the Indians who taught "corn planting" to early settlers; others cite Booker T. Washington's efforts at Tuskegee Institute in preparing Negro teachers. While these, the Land-Grant Act, and other historical events are indicative of early interest, the first two decades of this century are regarded as the formative years of teacher education in agriculture. It is to the conditions, institutions, and problems of this period that we look for (1) forces which influenced the development of, and shaped the course for, teacher education in agriculture; (2) the parent institutions which fostered and nurtured the fledgling profession; and (3) the special provisions leading to regularized status.

Teacher education in agriculture functions as an instrument of social policy. It is not an end itself. Its origins, therefore, may be expected to reflect socially determined needs (1.2). Furthermore, its establishment may be expected to relate to institutions and agencies concerned with similar purposes and problems. On the basis of these assumptions, it is therefore pertinent to review briefly agricultural developments, teacher education, agriculture colleges, and agriculture teaching in the public schools of this period.

Times and Institutions, 1900-1919

America in the period of 1900-1919 was a rapidly developing nation. Land for agricultural production was abundant and relatively low in cost. Farms were viewed as natural resources which could be more intelligently utilized. This, it was reasoned, would reward farmers and aid in maintaining low prices of food for the rapidly growing urban population. It was a favorable period for farmers; the years 1910-1914 were used for a long time as a base period in computing a price index for agriculture.

Science in this period achieved notable gains both in development of principles and in applications to agriculture. Advances in biological and physical sciences made possible a more definitive body of content in agriculture. The sciences, generally held in low esteem by older universities, were favored in the land-grant colleges. Research in the application of science to agriculture came to be accepted as the foundation for agricultural knowledge.

These were years of expansion in public education. Compulsory attendance laws were extended, and the school year was lengthened. High school enrollment increased fourfold. The percentage of youth aged 14 to 17 in school doubled from 1916 to 1920. This rapid growth in secondary education was accompanied by growing concern for its "proper function." A new consensus as to function was attained in this period. Citizens' and educational leaders' demands for more appropriate secondary education "found full expression in the profoundly significant work of the Commission on the Reorganization of

Secondary Education (1.3)." The Commission's report included vocation as one of its Seven Cardinal Principles.

The advocates of vocational subjects and activities in the curriculum gained strength from this action. Some professional educators gave considerable leadership to the movement in face of resistance from those who supported liberal education in the traditional sense. It was to these educators a way of reaching the underprivileged and disadvantaged; it was moving the public school toward the democratic ideal (1.4).

Land-Grant Colleges, 1900-1919

Land-grant colleges and universities advanced in stature during the period. The addition of the agricultural experiment stations and extension service broadened this influence greatly (1.5). Yet, judged by contemporary standards, the land-grant colleges too were in a transition period.

The Land-Grant Act was signed by President Lincoln in 1862. This historic and highly significant piece of legislation was an expression of popular will and needs, as reflected in the resolutions of the Illinois Farmers Convention of 1852, quoted by Keer.

1. As representatives of the industrial classes, including all cultivators of the soil, artisans, mechanics and merchants, we desire the same privileges and advantages for ourselves and our posterity in each of our several pursuits and callings, as our professional brethren enjoy in theirs.

2. The institutions originally and primarily designated to meet the wants of the professional classes as such cannot, in the nature of things, meet ours, any more than the institutions we desire to establish for ourselves, meet theirs.

3. Immediate measures should be taken for the establishment of a university, in the State of Illinois, expressly to meet these felt wants of each and all of the industrial classes of our state (1.6).

The Act included the following statement of purpose which is indicative of the broad vision of education advanced by Senator Morrill and other leaders of his day:

> . . . to promote the liberal and practical education of industrial classes in the several pursuits and professions in life.

Deans and faculty members in colleges of agriculture assisted in promoting agricultural education in elementary and secondary schools. They also assisted in aiding the normal schools to develop courses in agriculture for teachers. Much inservice work was done with teachers institutes and summer sessions and with materials dealing with nature study and agriculture. Booker

T. Washington's work at Tuskegee is widely known. Davenport of Illinois, Holden of Iowa, French of Michigan, Works of Cornell, and Vivian of Ohio were nationally recognized agricultural educators who promoted agricultural instruction in public schools. Elder states, "An outstanding accomplishment of the land-grant colleges is the development of ways to carry the new knowledge in agriculture to the people who can use it best (1.7)." Lord, in reference to his attendance at the "first agricultural high school" in 1908, states that the teacher-principal came as a result of "direct stimulus" of Dean Liberty Hyde Bailey of Cornell University (1.8).

Agricultural college faculty members, as well as professional educators, were active leaders in associations which contributed to the expansion of agricultural teaching. The following are some of the associations, with the dates of organization or major reports given.

American Association for the Advancement of Agricultural Teaching (1911)

Association of Agricultural Colleges (1897)

Country Life Commission (1908)

National Education Association (Committee Action of 1903-1907)

National Society for the Promotion of Industrial Education (1906)

National Society for the Scientific Study of Education (Yearbook 1912)

Nature Study Society (1908)

N.E.A. Department of Rural and Agricultural Education (1906)

The School Gardening Association of America (1911)

The U.S. Department of Agriculture became actively involved in the movement and published a bulletin entitled The Teaching of Agriculture in 1905 (1.9). It was not until 1908 that the U.S. Bureau of Education published Bulletin No. 1, The Training of Persons to Teach Agriculture, written by Liberty Hyde Bailey. Although land-grant colleges and departments of agriculture appear to have played leading roles in promoting instruction in agriculture, the contributions of teachers colleges, professors of education, state superintendents of public instruction, and other educators, as previously noted, were especially significant.

Education of Agriculture Teachers, 1900-1919

The movement to introduce agriculture into institutions other than colleges of agriculture gained momentum in this period. By 1915-16, agriculture was offered in some 160 private secondary schools in 37 states and in more than 3,000 public secondary schools as well as in normal schools (1.10). The rapid expansion of offerings had ". . . far outrun the ability of the colleges and normal schools to supply an adequate number of teachers trained in agriculture (1.11)."

In 1906, teacher education in agriculture reported at the land-grant institutions included normal courses, summer sessions, and programs in departments of education. By 1910, 46 agriculture colleges reported teacher-training work in agriculture (1.12). However, this work was chiefly inservice in nature and evidently was not regarded as constituting a teacher-training program. (See Table 1.1.)

TABLE 1.1--States initiating teacher education in agriculture, by year, prior to and subsequent to the passage of the Smith-Hughes Act

College Year	States Initiating	Total Cumulative	College Year	Number of States Initiating		Total Cumulative	
				White	Negro	White	Negro
1908-09	1	1	National Vocational Education Act passed				
1909-10	2	3					
1910-11	3	6					
1911-12	1	7	1917-18	8	3	27	3
1912-13	2	9	1918-19	13	4	40	7
1913-14	4	13	1919-20	6	3	46	10
1914-15	4	17	1920-21	1	2	47	12
1915-16	1	18	1921-22	1	--	48	--
1916-17	1	19	1937-38	--	1	--	13

Source: Federal Board for Vocational Education Bulletin No. 100, Agricultural Series No. 23, May 1925, p. 9; Bulletin No. 111, Agricultural Series No. 28, May 1926, pp. 34-35; Bulletin No. 217, Agricultural Series No. 55, 1943, pp. 5-6.

The broadened purposes of secondary education facilitated expansion of offerings, including agriculture, to a point where special provisions were required to ensure a supply of teachers. Support was given for establishing teacher education in agriculture on a regularized and continuing basis through the Smith-Hughes Act of 1917. It required states that received federal funds to allocate some resources to the education of teachers of agriculture. In practically all states, the land-grant colleges or universities were designated as institutions in which this work would be undertaken (1.13). By 1919, most of the states reported definite provisions for the education of teachers of agriculture, as indicated in Table 1.1.

Shortly thereafter, all the states, plus Hawaii and Puerto Rico, had initiated agricultural teacher education programs. In several states, more than one program was initiated until, by 1967, there were 76 active teacher education programs. The programs in Hawaii and Maine and at Kentucky State College had become inactive by 1967, but the program at Rutgers, which had been inactive for some time, was reactivated. In 1980, there were 78 institutions with programs distributed over 48 states. Sixteen states had two or more institutions with programs. The program in Puerto Rico had increased to 79. Only Hawaii and Alaska did not have agricultural teacher education programs.

The provisions of the Smith-Hughes Act pertaining to the development of teacher education in agriculture were summarized by Swanson (1.14) as follows:

1. Provides for a permanent and continuing appropriation for the preparation of teachers, supervisors, and directors of agricultural subjects.

2. Makes it mandatory for each State accepting the provisions of the act to designate or create a State board with power to cooperate with the Federal Board of Vocational Education, and stipulates that States must use the minimum amount appropriated for the training of teachers in order to secure the other benefits of the act. Thus added emphasis is given to the importance of adequate provision being made in each State for the preparation of teachers.

3. Requires State boards to prepare state plans for vocational education which shall include plans for training of teachers.

4. Stipulates that funds appropriated for the preparation of teachers shall be matched dollar for dollar by the State or local community or both and used for the maintenance of such training.

5. Provides that the training of vocational teachers shall be under the supervision of the State board for vocational education, sets up specifications which must be adhered to in training programs, and provides that not more than 60 per cent nor less than 20 per cent of the money appropriated to any State for any year shall be expended for the training of teachers of agriculture.

6. Prohibits the use of funds appropriated under the act for "the purchase, erection, preservation, or repair of any building or buildings or equipment or for the purchases or rentals of lands, or for the support of any religious or privately owned or conducted school or college."

Teacher educators in agriculture staffing programs in the newly designated institutions represented a widely diverse background of experience and education. They included classical scholars, drafted professors of agricultural subjects, and secondary school teachers of agriculture; a minority of them possessed graduate degrees with emphasis in psychology or education. Their first efforts in developing programs presented an equally varied picture, a function in part, it should be recognized, of the difference in situations as well as in preparation.

The establishment of periodic national and regional conferences of supervisors and teacher educators by the Federal Board for Vocational Education contributed to the development of a cohesive group with unified purposes (1.15). True's comments on the effect of teacher education upon the land-grant colleges is worth reviewing in order that the interaction of various movements may be better understood.

> The assumption of the duty of training teachers for the secondary schools has affected the agricultural colleges favorably in several ways. It has greatly broadened the interest of the authorities and teachers in the problems of agricultural education and the application of pedagogical principles of the teaching of agriculture. It has opened a new vocational outlet for a considerable number of graduates from the agricultural courses of these colleges. It has given these colleges more prominence in the thought of the pupils in many high schools and brought a considerable number of them to the colleges for long or short courses. It has fundamentally affected the relation of these colleges to the public-school system of the several states and made them more fully an essential part of this system. Since the United States has only begun to develop a comprehensive system of vocational education, it may be expected that with the accelerated progress which such education will make the colleges standing at the head of the agricultural division of this system will have an increasingly important part to play in its development and maintenance (1.16).

Some Conclusions

Each specialized occupational group is formed under pressure of social needs. Its initial responsibilities are in part an extension of those of institutions which have been active in securing recognition of the need.

In the special case of teacher education in agriculture, one may conclude:

1. The convergence of forces represented in expanding the science of agriculture, increasing abundant agricultural resources, and broadening the purposes in the public secondary school contributed to a rapid expansion of the teaching of agriculture in the period of 1900-1919.

2. The land-grant colleges, particularly their agricultural faculties, were advocates of the extension of agricultural education of less than college level in public schools.

3. The expansion of demands for agricultural education and the support of many professional educational associations, the land-grant colleges of agriculture, and the U.S. Department of Agriculture contributed to the passage of legislation giving teacher education in agriculture a regularized status in the land-grant colleges.

4. By 1920, considerable progress had been made in establishing a cohesive movement in teacher education in agriculture from what had been disparate and diverse efforts of many institutions.

5. The prime mission became that of preparing teachers of vocational agriculture to aid in democratizing the public schools and advancing agriculture.

Orientation and Outlook

From the "formative years" of 1900-1919, teacher education in agriculture emerged with a pioneering and progressive outlook. Its orientation was toward the enhancement of democratic institutions committed to improvement of education for all citizens. These views were widely shared by leaders in the land-grant colleges and in public education, with which teacher education in agriculture was closely identified. In general, the land-grant colleges of agriculture had assumed or accepted major responsibility for the preparation of teachers of vocational agriculture, and this responsibility had gained assurance of support from the federal government. There were prospects of expanding demands for teachers, the challenge of influencing the character of agricultural instruction in secondary schools, plus a sense of involvement in the cause of advancing scientific farming. With this sense of mission, teacher education in agriculture entered a period of maturation, from 1920 to the present period.

How firmly did the early orientations and outlooks persist in the years of "maturation"? What forces acted to modify them, and what were the emergent orientations and outlooks of teacher education in agriculture? These questions should direct our attention specifically to the impact of these changes on the preparation of teachers of agriculture. In large measure, the orientation and outlook of teacher educators are the roots from which grow the purposes and processes of future teachers. They are the basis for defining the proper role of the profession, including the "model teacher of agriculture" and the "model program" for which he/she is responsible.

Professional Involvement and Status

Teacher educators in agriculture developed many occupationally related associations. They were members of university faculties and co-workers in departments. They were college professors, an occupational category which ranked high in prestige and income. They were identified with public education and held active membership in a variety of professional education associations. Further, many were members of honorary, scientific, and professional associations primarily associated with agriculture. Thus, teacher education in agriculture developed as a sub-discipline with unique ties to both education and agriculture. It was subject to general sanctions of professions and to the specific professional codes governing education. Within this framework, its own orientation and outlook, which guided decisions and actions with regard to the education of teachers of agriculture, evolved.

Teacher education in agriculture was initiated and nurtured during a period of great expansion of professional occupations. From 1900 to 1960, the labor force engaged in professional occupations increased from 4.3 to 11.2 percent. Teachers became the largest professional group, representing over one-fourth of the total professional employment. Teachers of agriculture in public schools increased from practically none to about 10,000 in this same period.

Professional employment increased not only in numbers but also in terms of prestige and status. The concept of a profession, including its responsibilities and rights, reflects the foundation common to the professions but which is not characteristic of other specialized occupations. Flexner presented a list of distinguishing traits of the medical profession in 1915. More recent works by Greenwood (1.17), McGlothlin (1.18), and Stinnett (1.19) relate these criteria to the professions and specifically to teaching. The following paragraph summarizes their views as to qualities which mark an occupation as professional.

A profession involves a personal commitment to serve society above self, even though a fee or profit is the individual's right. It requires capacity to deal with complex and variable situations involving the exercise of choice and judgment. Extensive educational preparation, primarily intellectual, is required to enter the profession and to master the knowledge and techniques essential to its practice. Yet, a practical outlook is required, since this special knowledge must be used or translated into action for the good of humanity. This ideal of service generally enables the associations representing professions to obtain legal support for standards of licensing or certifying persons desiring to enter the profession.

These generalizations constitute a common code which is presumed to orient the purposes or aims in many professions and professional schools, in particular. The provision of competent new members, adequate in numbers to serve the needs of society, is given by McGlothlin as a prime purpose of a profession. Additional purposes which are characteristic of the professions are advancement of knowledge through research, protection of members and practice, and unification of members in rendering improved service.

Teacher Certification

With society growing more complex, states exercised many regulatory functions including the licensing of doctors, lawyers, and teachers. In 1921, 15 states had developed centralized licensing of teachers. With limited exceptions (delegation of authority to cities), by 1965 all states required teaching certificates (1.20). In many states, a close degree of cooperation developed with the institutions preparing teachers. These institutions were further accredited by one or more professional organizations or associations.

In the early days of agricultural education, a certificate was not generally required to teach agriculture. The state plans under the Smith-Hughes

Act required each state to indicate qualifications for teachers of agriculture. As the general movement to require teaching certificates was extended, it included teachers of agriculture.

Most states, in addition to requiring general education and education in teaching fields, mandated work in professional education which comprised special methods of teaching and student teaching. The latter consisted of courses in which teacher educators in agriculture had a vested interest, and it may be presumed that they gave their backing to including them in certification requirements. The professional faculties of teacher education institutions, the state certification bureaus, and the professional organizations involved in accreditation were part of what was referred to as the "educational establishment." Conant focused attention upon the relation of the educational establishment to certification and recommended that the state certification requirements be modified by granting more autonomy to institutions (1.21). Certification regulations affect teacher education programs. They may in effect set the curriculum (1.22). In those states whose policies are preventing institutions from improving teacher education programs, forces concerned with teacher education should work towards improved policies of teacher certification.

Teacher Education Institutions

This century saw great advances in the quality of teacher education programs. It became a college function. Two- and four-year normal schools became teachers colleges; most of which are now known as state colleges. With few exceptions liberal arts colleges, public and private universities, and state colleges are now engaged in teacher education (1.23). There were over 1,200 of these institutions in 1965, including the land-grant colleges, most of which became universities after 1920 (1.24).

Public colleges and universities provided an increasing proportion of new teachers, reaching 70 percent by 1965. Land-grant colleges and universities, considered in terms of numbers of graduate and undergraduate degrees in education, became major suppliers of new teachers, especially for secondary schools.

The development of teacher education in agriculture was influenced by the growth of teacher education in universities. Policies and programs were increasingly developed on an institutional basis. Resources of new knowledge and materials were developed and utilized for teachers; these involved cooperative relations of staff members from many departments.

In 1920, teacher education in agriculture was one of the few areas in which teachers were prepared in land-grant colleges. It was in a relatively strong position with special support and an expanding demand for teachers. Its rate of growth after this period was slow in comparison with that of land-grant universities and other teacher education institutions. In 1966, the directory of teacher educators in agriculture (1.24) listed 76 institutions conducting teacher education in agriculture programs. The institutions employed about 260 teacher educators in agriculture, but 33 institutions employed only 1 or 2 persons in this capacity. The 1979-80 directory lists 78 institutions, plus Puerto Rico, with programs and 376 teacher educators.

This increase of 116 (45 percent) during the period 1966-1979 establishes this period as a time of significant growth.

Identification with Local Programs

Teacher educators in agriculture had extensive contacts with secondary schools. These contacts resulted from promotional, placement, and training activities. The content of agricultural education was not well established, and study of local programs was a common means used for the further development of content for professional courses. Many teacher educators served as itinerant teacher trainers, and in the early 1920's nearly one-half of the states had only one position known as "supervisor-teacher trainer." In addition, it was not unusual for the supervisory staff members to be stationed at the land-grant colleges of agriculture. Thus, the situation favored, if not required, a close relationship with local programs.

The policies of the Federal Board for Vocational Education were revised in 1922 to make it possible to use teacher-training funds under the Smith-Hughes Act for inservice teacher training and supervision. This change was criticized by the Advisory Committee on Education as weakening teacher education (1.25). It did, in many states, mean that teacher educators were expected to spend substantial time providing inservice education, thus increasing contacts with members of the profession in the schools.

The involvement with local programs inclined teacher educators in agriculture towards giving the local programs a central place in the scheme of things. The orientation was toward local programs, based on the proposition that "as they improve so will vocational education in agriculture." Their improvement was viewed as dependent almost completely upon local effort under the leadership of the teacher of agriculture. The identification with local programs enabled the continued development of concepts as to the proper role of a teacher of agriculture and the formulation of generalizations regarding the organization and conduct of local programs of vocational agriculture. It was, in part, an extension of the pioneer outlook--a desire to be close to the frontier. But as frontiers are conquered, new ones present challenges.

The centrality of a local community was challenged by the need for a broader orientation even before the 1950's. Emergencies generated by depression, droughts, wars, and drastic changes in agricultural policies were shifting the focus of agricultural educators even before Sputnik (1957). Regional, national, and world views competed for attention.

Dedication to Production Agriculture

Colleges of agriculture were committed to "making two blades of grass grow where only one grew before." The science and mechanization of farm production enabled production to increase throughout this period, but the rate of increase was further accelerated in the late 1950's. Agricultural educators on all fronts were involved in the movement to increase production efficiency. For some, the increased efficiency appeared as a way by which low incomes of those in farming could be raised; for others, the very challenge of discovery and the application of science were sufficient.

Teacher educators in agriculture joined in the drive to advance agricultural efficiency. To the prevailing thrust from the college of agriculture was added the impetus of federal and state policies growing from the Smith-Hughes Act, with emphasis on developing proficiency in farming.

The resultant orientation of teacher educators centered on the position of farm owner-operator. The clientele of teachers of agriculture were perceived from this orientation. Planning, conducting, and evaluating local instructional programs by teacher educators was closely related to the needs of these clients (1.26). Experience in farm production, preferably as owner or operator, was viewed almost as an essential for those admitted to preparation for teaching agriculture and was usually found in certification regulations.

Disenchantment with this orientation became widespread prior to 1960. Farm production in America, both in totals and per farmer, became the envy of the world; yet there was continued decline in the number of farmers and in opportunities for youth to become established in farming. Employment, on the other hand, increased in those industries which provided services to farmers and processed and distributed the products of the farm.

A new orientation was required, not to displace fully that of production agriculture, but rather to supplement it, if agriculture was to retain its place in the secondary school curriculum. The Vocational Education Act of 1963 legalized the new orientation which had been advanced by teacher educators in agriculture and by others for some years. A comparison of published statements of objectives of vocational agriculture of 1940 and 1966 will afford further evidence of the change in orientation reflecting a broadening of social and occupational perspective. Additionally, a broadened curriculum is reflected in the 1977 Standards for Quality Programs in Agricultural/Agribusiness Education (1.27) which lists standards for seven program areas: production agriculture; ornamental horticulture; agricultural mechanics; agricultural products; agricultural supplies and services; forestry; and agricultural (natural) resources.

Science and Education

The movement to utilize the methods and findings of science in education was initiated even before 1900. Early leaders in this movement, including Dewey, Kilpatrick, and James, are cited by Stimson for their influence on agricultural education (1.28). Teacher educators, as has been indicated, were proponents of the movement to make farming scientific. The physical and biological sciences were regarded as the foundations for agriculture and, judging by curricula, for agricultural education. Later, there was recognition of the importance of the social, or behavioral, sciences. The increase in knowledge of the behavioral sciences within this period was substantial. In 1960, the Encyclopedia of Research in Education comprised a book of 1,564 pages. A Handbook of Research on Teaching, published in 1963, included over 1,200 pages. The Second Handbook of Research on Teaching, published in 1973, contained 1,400 pages.

Beginning in the middle part of this period, agricultural teacher educators instituted regional research conferences, which were encouraged by the U.S. Office of Education. There is reason to believe that these conferences marked an awakening to the necessity for more attention to the behavioral sciences. In part, this interest reflected a concern for research, but it also involved teacher education. Cooperative research organized through these conferences resulted in studies of training needs, recruitment procedures, teacher retention, public relations, and other problems of direct concern to the education of teachers of agriculture. Regional research conferences and seminars have been continued to the present time.

Toward New Orientation and Outlook

It is recognized that not all teacher educators have the same orientation and outlook. These efforts at generalizations are recognized as a partial and incomplete basis upon which to begin a more critical examination of the assumptions which underlie decisions and actions in teacher education in agriculture.

1. There is a shift in major orientation from that of agriculture to education, with a focus on professional responsibilities of the teacher, brought about in part by the influence of teacher certification and institutional accreditation.

2. Problems and issues are viewed increasingly from a state or national reference point rather than from a local viewpoint.

3. The behavioral sciences are viewed with increasing favor for their possible contribution to improvement of teacher education which tomorrow's world requires.

4. There is a broadened outlook with regard to contributions which teachers can make and the persons they can serve.

5. The new outlook favors graduate teacher education over visitation and similar inservice teacher education activities.

6. The focus of vocational agriculture curricula has shifted from production agriculture to off-farm businesses and industries providing services to farmers and processing/distributing farm products.

7. Regional and national efforts by professional organizations representing agricultural teacher educators, state supervisors, and teachers have tended to produce more uniform quality and content in teacher education programs.

Organization and Administration

The 1960's were the beginning of a period of reorganization in higher education. Teachers colleges became state colleges; plans for coordinating higher education were developed in many states; drastic revisions of teacher certification regulations took place; and colleges developed innovative patterns of organization. Some new institutions did not provide for "separate professional schools or even separately designated professional curricula (1.29)." This ferment and change in organization of higher education was destined to continue into the 1970's. In this period the organization and administration of teacher education in agriculture experienced varied opportunities and challenges.

Many of the forces for change in education are focused upon the organization and administration of teacher education. There are strong pressures from academic professors and lay citizens for a "redirection of public authority" in teacher education (1.30). A major point of contention centers on the degree of autonomy which should reside in the "educational establishment," including professors of education and departments of education. At the heart of the matter is a belief that teachers should be better educated in academic disciplines which, it is asserted, have been sacrificed to the so-called "less educative" course requirements in professional education.

The current emphasis in teacher preparation on competency/performance-based teacher education originated, in large measure, from public sentiment for greater accountability in public education. The move in several states toward statewide teacher tests and licenses also reflects this sentiment.

Teacher educators differ in their views of the central features in teacher education. Stratemeyer identified three central areas of teaching: scholarship, direct experience, and the interrelationship of research and teaching (1.31). To these might well be added student recruitment and admission, in view of previous and current shortages of teachers of agriculture and the profession's obligations to provide an adequate supply of teachers. The focus, therefore, is on forces which act strongly upon these central concerns.

Current Situations

Table 1.2 lists institutions that prepare teachers of agriculture, the number of agricultural teacher education faculty members in each, and the school or college in which each agricultural teacher education department is located. The number of faculty members listed is based on the Directory of Agricultural Teacher Educators, 1979-1980, issued by the American Association of Teacher Educators in Agriculture. The figures include, in addition to those faculty members in departments of agricultural education, persons outside the departments who had some responsibility for providing professional education courses--principally instructors of agricultural mechanics and instructors of methods of teaching agricultural mechanics who were found in departments of agricultural engineering.

In the cases of North Carolina State University and The Ohio State University, several individuals holding professorial appointments in the departments of agricultural education spent all or portions of their time in the vocational education research centers located on those campuses.

In the land-grant universities, about half of the agricultural teacher education departments were located in colleges of agriculture and about half in colleges of education. There does not seem to be a clear answer to the question: "In which college should agricultural teacher education be

TABLE 1.2--Number of faculty in agricultural education and their academic affiliations in institutions offering agricultural teacher education, 1979-80

Institution	No. of Faculty	School or College
Auburn University (Ala.)	5	Educ.
Alabama Agricultural & Mechanical University	5	(b)
Tuskegee Institute (Ala.)	3	Agri.
University of Arizona	4	Agri.
University of Arkansas	5	Educ.
Arkansas State University	2	Agri.
California State University	3	Agri.
University of California	2	Agri.
California Polytechnic State University	8	Agri.
Colorado State University	3	Educ.
University of Connecticut	1	Educ.
University of Delaware	2	Agri.
Delaware State College	1	(b)
University of Florida	14	Educ.
Florida Agricultural & Mechanical University	1	(b)
University of Georgia	6	Educ.
Fort Valley State College	1	Agri.
University of Hawaii	--	(a)
University of Idaho	3	Agri.
University of Illinois	6	(c), Educ.
Western Illinois University	3	Educ.
Southern Illinois University	9	(d)
Illinois State University	4	(b)
Purdue University (Ind.)	4	Educ.
Iowa State University	11	Agri.
Kansas State University	7	Educ.
University of Kentucky	6	Educ.
Morehead State University (Ky.)	1	Educ.
Murray State University (Ky.)	2	Agri.
Western Kentucky University	2	(d)
Louisiana Tech University	1	(b)
University of Southwestern Louisiana	1	Agri.
Southern University (La.)	1	(b)

(Continued)

TABLE 1.2 (Continued)

Institution	No. of Faculty	School or College	Institution	No. of Faculty	School or College
University of Maine	1	(a)	North Dakota State University	3	Agri.
University of Maryland	6*	Agri.	The Ohio State University	17*	Agri.
University of Maryland - Eastern Shore	3	Agri.	The Ohio State University Agricultural Technical Institute	1	(e)
University of Massachusetts	2	Educ.	Kent State University	2	Educ.
Michigan State University	6	(d)	Oklahoma State University	7	Agri.
University of Minnesota	9*	Educ.	Oregon State University	4	Educ.
Mississippi State University	19	Agri.	The Pennsylvania State University	19*	Agri.
University of Missouri	7	Educ.	University of Puerto Rico	6	Agri.
University of Montana	7	Agri.	University of Rhode Island	2	Agri.
The University of Nebraska	9	Agri.	Clemson University (S.C.)	5	Educ.
University of Nevada	2	Agri.	South Dakota State University	3	Educ.
University of New Hampshire	2*	Agri.	University of Tennessee	1	Educ.
Rutgers The State University of New Jersey	4*	Educ.	Tennessee State University	9*	Agri.
New Mexico State University	6	Agri.	Middle Tennessee State University	1	Educ.
Cornell University (N.Y.)	6	Agri.	Texas A&M University	14	Agri.
North Carolina State University	3	Educ.	East Texas State University	3	(b)
North Carolina Agricultural & Technical University	5	Agri.	Sam Houston State University	1	(b)
			Southwest Texas State University	1	(b)

(Continued)

TABLE 1.2 (Continued)

Institution	No. of Faculty	School or College
Texas Technological University	4	Agri.
Texas A&I University	1	Educ.
Austin State University	1	Educ.
Tarleton State University (Tex.)	3	Educ.
Prairie View A&M University (Tex.)	1	Agri.
Utah State University	4	Agri.
University of Vermont	4*	Agri.
Virginia Polytechnic Institute and State University	13*	Educ.
Virginia State University	3	Agri.
West Virginia University	6*	Agri.
The University of Wisconsin	2*	Agri.
Wisconsin State University - Platteville	3	Agri.
Wisconsin State University - River Falls	3	Agri.
University of Wyoming	2	Educ.

*Includes individuals with responsibilities elsewhere in the institution.

(a) Department inactive.

(b) In a combined department of agriculture.

(c) In addition, there are five staff persons in the Illinois Vocational Agriculture Service, which is located in the college of agriculture.

(d) Joint location in colleges of agriculture and education.

(e) Two-year institution.

located?" Generally, deans of agriculture feel that agricultural teacher education should be located in their department so that agricultural education students can be in close contact with their subject matter professors. In addition, in most institutions, agricultural education students enroll in colleges of agriculture, regardless of which college the faculty in agricultural teacher education is administratively located. Deans of education usually feel that professors of agricultural education should be administratively located in colleges of education with other professors of education.

While some of the non-land-grant institutions are faced with the same division, in many of them there is no school or college of agriculture. Various forms of organization exist, as shown in Table 1.2. The departments may be combined with a department of agriculture or may exist as separate departments.

With respect to titles, in all the land-grant institutions and most other institutions, agricultural teacher educators bear the title of instructor, assistant professor, associate professor, or professor.

Those agricultural teacher educators who favor location in the college of agriculture may have one or more faculty members with experiment station appointments. Thus, part of their salaries are paid from that source, and they are provided time and funds for the conduct of research. According to the 1966-67 directory of Professional Workers in State Agricultural Experiment Stations and Other Cooperating State Institutions, agricultural teacher educators in nine institutions held experiment station appointments (1.32).

Also, the establishment of research coordinating units in vocational education on campuses in some states has made it possible for faculty members located in colleges of education to have portions of their time allocated to research.

In terms of training, agricultural teacher education personnel have about as much formal training as other professors in agriculture and in education. Excluding faculty members with the rank of instructor, who are usually temporary or who are engaged in part-time graduate study, almost all agricultural teacher educators hold the doctorate.

In spite of the variety of organizational patterns, there have been common elements which tend to give some singleness of character to agricultural teacher education in all institutions:

1. The federal government, through the U.S. Department of Education, has indirectly provided financial support for most of the positions.

2. The institutions were designated by state boards of vocational education, and the programs were subject to the teacher education provisions of the state plans for vocational education (1.33).

3. Teacher education in agriculture was usually established within an institution's framework and, in general, were bound to function within its policies and under its administration.

4. The development of cooperative relationships with secondary schools is essential to carrying out teacher education.

Professional Organization

It should be noted that teacher educators organized better to achieve their purposes. The development of regional research conferences, previously referred to, is one example. Teacher educators were leaders in initiating the Agricultural Education magazine in 1929. This professional publication has been a prime source of materials for undergraduate courses, as well as a valued reference for teachers. In 1932, teacher educators and others launched a project to collect and publish summaries of research in agricultural education. The first of the series of Summaries of Studies in Agricultural Education was published in 1935 as Vocational Education Bulletin No. 180, and this effort has continued to date. The research committee of the American Vocational Association assumed responsibility for this project. In 1959, teacher educators formed the American Association of Teacher Educators in Agriculture (AATEA), which is affiliated with the American Vocational Association (AVA).

The Journal of the American Association of Teacher Educators in Agriculture was first published by AATEA in 1960. It has evolved to the present time as a refereed journal for reporting research in agricultural education.

The first edition of Teacher Education in Agriculture and this second edition were compiled through the efforts of AATEA. AATEA has also taken leadership in the development of a number of publications in agricultural education and has worked cooperatively with teacher and state supervisor organizations in sponsoring conferences and promoting standards for the profession.

It was primarily through organized efforts of the profession that a national center for agricultural education was established at The Ohio State University. Its success clearly demonstrated the need for a larger unit in meeting needs of professional education. In 1965, its title became National Center for Vocational and Technical Education, thus representing a further broadening of scope and purpose. Its current title is National Center for Research in Vocational Education.

Professional organization activities contributed to important features in teacher education. They were especially significant in the case of teaching, scholarship, and research. The profession, through its professional organizations, is one of the most important forces for the continued increase in effectiveness of teacher education in agriculture.

Federal Influence

Early policies in vocational education admittedly did much to improve teacher education in agriculture. They focused attention on the needs for establishing standards with respect to selection, curricula, qualifications of personnel, student teaching, inservice education, and research related thereto (1.34). On the other hand, these policies contributed to a high degree of autonomy. The reimbursement policies rewarded institutions in which teacher educators devoted all their time to agricultural education. In themselves, these policies are not sufficient to explain the present pattern, since departments in which agricultural education is linked with other similar interest groups evolved in a number of institutions.

Recent policies in connection with the Vocational Education Act of 1963 and subsequent amendments may have an opposite effect. The identification of separate fields is less distinct. New occupational training programs will require combinations of competencies. The reorganization of vocational education at the national level reflects a multidisciplinary approach. The appropriations for research in vocational education are available for approved projects in almost any department of a university. It thus appears that the policies of the U.S. Office of Education since 1963 have favored a broader pattern of relationships than previously existed, a factor which may influence a change in departmental organization.

Bueke, reporting on the impact of the 1976 VEA Amendments, stated, ". . . it appears that the 1976 amendments have exerted substantial leverage on the vocational system in our sample states. . . . Federal goals and priorities strongly influence state activities (1.35)." In states with limited resources in particular, the impact of federal funds tied to goals and priorities seems to be a major influence at the state level.

Some priorities are emphasized by "set aside" funding, for example, 10 percent to vocational programs for disabled students in the 1976 VEA Amendments. Funds set aside may be spent only for the designated priority area.

State Controls

State boards of education control the amount of federal funds allocated to teacher education in agriculture. In several states they exercise control over the budget, expense accounts, and other fiscal matters. They may determine the number of positions to be allocated to an institution and specify certain types of inservice activities or even non-teacher education functions. As we have seen, the states, through certification regulations, may exercise a decisive influence on curricula. The influence of state boards' policies administered through their agents, in general, has reinforced the autonomy drive of teacher education in agriculture within institutions.

Checks and balances are required over all public agencies and institutions. The influence of state departments may be viewed in this light. It may also be argued that universities "lead the way to a greater society" and should be subject only to very broad and general regulations. Judging by the temper of the times, it is questionable if public colleges and universities (not necessarily departments) are likely to be granted increased autonomy in the organization and administration of their programs.

For several years prior to 1963, several states had written contracts or agreements between their state departments of education and their agricultural teacher education institutions. The 1963 Vocational Education Act (VEA) requires such an arrangement. The contract specifies the functions that the agricultural teacher education faculty and the institution will perform and the nature and extent of reimbursement to the institution to be provided by the state department of education. The completion of such an agreement by the two agencies is more likely to result in clarification of expectations of one another and avoidance of situations in which one is involved in activities which the other considers its responsibility.

The allocation of federal funding within states is determined by annual state plans for vocational and technical education. Federal approval of state plans on an annual basis is a contingency for funding of the plan. Thus, persons writing state plans exercise considerable control, and it is important for teacher education to be included as part of the plan. State advisory committees for vocational and technical education are required under VEA legislation. These advisory groups provide opportunities for input outside state departments of education.

Institutional Patterns

It seems natural to view the institution as central in considering organization and administration. The federal, state, and local influences are usually at least one step removed from those of the departments of universities and colleges.

Teacher education in agriculture, like other specialities, was generally added as a department and usually continued in that independent status. This continued accretion of functions and departments led to much concern on the part of college administrators. Administration is more difficult with larger numbers of units. Duplication of effort increases, and growing problems in maintaining communication between departments make it difficult to develop cooperation (1.36). These and other disadvantages in the existing pattern have implications for teacher education in agriculture. These disadvantages may be close to the administrator's role--but are they not also related to improvement of teacher education? One's responses, of course, would be dependent upon that person's conception of the role of the teacher and how best to prepare him/her.

Greatly increased enrollments during the 1960's and the early 1970's added to the problems of administration of land-grant institutions. The increased number of applicants, in some states, was parallel with raising requirements for admission and/or graduation. Certainly, after Sputnik, colleges and universities devoted major efforts to advancing the level of instruction. Increased standards, combined with a decline in the relative attractiveness of agriculture, made for difficulties in obtaining sufficient teaching candidates--a problem which exists today. There was, of course, a general problem, as professional opportunities were expanded and requirements were raised. In the case of teacher education in agriculture, continuing problems of obtaining an adequate supply of teaching candidates at the desired levels of academic ability may be anticipated.

Teacher education in agriculture in a few institutions developed a cooperative relationship with extension education. This led to a broadened departmental structure. Other institutions combined all vocational teacher education into one department, although this pattern was used in some institutions from the beginning. Smaller states have proposed interstate pacts which would result in pooling of efforts to conduct teacher education in agriculture at one institution. Teacher education in agriculture participated in one foundation-sponsored program involving a number of institutions in developing a Master of Arts in Teaching program. These and other developments reflected changes in organization.

The current decreasing school age population, along with general sentiment for greater accountability in public education, has resulted in program cutbacks and reorganization. The continuing shortage of agriculture teachers may be offset by reduced demand due to declining enrollments. It appears that the 1980's will be a period of limited, if any, growth in teacher education programs. Federal support for vocational education is being reduced, and state appropriations are not keeping pace with inflation. New organization patterns and cooperative efforts targeted toward cost-effective use of resources can be anticipated for the coming decade.

The growing value attached to research by institutions and the decreased availability of funds in vocational education are forces influencing central concerns of teacher education in agriculture. Research activities contribute to institutional and departmental prestige and are a prime requisite to individual success in the academic marketplace (1.37). The probable gains from more emphasis on research outweigh dangers to the central concerns, if a sound relationship can be maintained between teaching and research activities. Again, it appears easier to maintain this relationship as well as to enhance teaching-scholarship when several individuals are included in the departmental unit. Furthermore, the interdisciplinary approach to teacher education may be facilitated as these forces gain strength.

Current trends are for research funding to be targeted on state and federal priorities, such as for disabled individuals and for basic skills development. Research efforts in teacher education can be predicted to reflect these priorities.

Direct Experience

Most professional training involves some type of direct experience which is, in part, organized and administered by the professional school. Questions of control over the direct experience phase of professional education have been raised in medicine and other areas as well as in education (1.38). The direct experience, in the case of teacher education, is organized for varying lengths of time, ranging from a few weeks to a full year. Most teachers and educational leaders, including Conant, regard it as potentially a most valuable feature of the program. Since it is treated in depth in a later chapter, our attention is restricted to a brief view involving administration and organization.

The growth of teacher education at universities and colleges was conducive to the development of controls over all direct experience, or student teaching. This led to the creation of a position of director of student teaching in many institutions. General policies developed which applied to most areas of teacher education. Certification and accreditation movements also favored a common policy (1.39).

The growing numbers of students, increasing lengths of experience, and other forces caused school administrators and cooperating teachers concern. They, too, became involved in policy determination as it related to the selection and payment of cooperating teachers, the selection of student teachers, and the nature and extent of experience and evaluation.

Direct experience in agriculture is provided within some frame of organization involving the total university, secondary schools, and members of the profession. It is the phase of the program most directly involved in an interdepartmental type of organization and most subject to influence of forces in the profession. It is a phase of the program requiring continued innovation in patterns of organization and administration.

The current movement toward competency/performance-based teacher education emphasizes evaluation of competencies in actual school settings. This movement has resulted in increased importance and expectations for field experience. (The competency/performance movement is addressed in Chapter 3.)

Additionally, a trend for extended field experience is emerging. The New York State Board of Regents has included a one-year internship as a part of the new licensing requirements for all teachers.

The current trends indicate that field experience will be an area of emphasis and change in teacher education programs during the 1980's.

The Contemporary Period

The period of 1967-1981 was characterized by expansion followed by contraction in higher education. These changes reflected the enrollments in higher education as the college age population peaked in the mid-1970's and then began an extended period of decline, which is projected to continue well beyond the present date. At the same time, agricultural education experienced a 45 percent increase in teacher education staff (see Table 1.2), and the profession engaged in significant efforts to plan for the future and to improve program quality.

Planning for the Future

In 1967, the agricultural education profession began planning for the 1970's and beyond through the Committee on Agricultural Education of the Commission on Education in Agriculture and Natural Resources (1.40). The report of this committee, published by the American Vocational Association in July 1971 represented the priorities and philosophy of the profession at that time.

In July 1980, a national conference on agricultural education in the 1980's was held in Kansas City, Missouri. The conference was sponsored jointly by the agricultural education profession and the U.S. Office of Education. Major leadership for the conference was provided by teacher educators. These planning efforts indicate a maturation of teacher education in agriculture and a recognition of the need for national-level coordinated planning for all agricultural education.

Development of Standards

Early efforts to develop standards for teacher education in agriculture began in 1951 as a joint effort of the agricultural section of the American Vocational Association and the Association of Land-Grant Colleges and Universities to establish criteria for institutions preparing teachers of agriculture. Based on this work, the Teacher Education Committee of the agricultural section began further study, which culminated in Guiding Principles for Pre-service Training of Teachers of Vocational Agriculture (1.41). These principles were approved at the AVA convention in 1961 and published in 1962.

In 1968, a group of agricultural teacher educators and supervisors prepared the document Guidelines for Agricultural Education. In November 1969, the American Association of Colleges for Teacher Education (AACTE) published Recommended Standards for Teacher Education (1.42). Using these standards as a starting point, the American Association of Teacher Educators in Agriculture Guidelines Committee subsequently developed guidelines for teacher education in agriculture (1.43). The Teacher Education in Agriculture Guidelines report was approved at the AVA convention in 1971.

The aforementioned efforts, in conjunction with three national conferences during the period, provided a foundation for the 1976-77 project to validate and disseminate standards for quality agricultural/agribusiness education (1.44). The standards are placed in 11 categories, one of which is teacher education. A national conference on the standards was held in Kansas City, Missouri, and the standards developed were disseminated for validation in all 50 states through regional meetings in 1977. The U.S. Office of Education-funded project was conducted by Iowa State University. Implementation of the standards is on-going at the present time.

A Philosophy of Vocational Agricultural Education

Many of the earlier developed guidelines, principles, and standards for agricultural education contained statements representing a philosophical point of view. However, until the Dougan statement in 1975 (1.45), no statement as such had been endorsed by the agricultural education profession on a national basis. His statement, approved by the Agricultural Education Division of the American Vocational Association in 1975, represents an initial effort toward consensus on a philosophical position. The following are the key concepts in the Dougan statement:

> The philosophy of vocational agricultural education provides the general explanation for the program. This general rationale is the base from which the program is developed.
>
> The development of a program of vocational agricultural education requires a series of standard concepts that provide lasting stability and direction, and which are compatible with the philosophical foundation. These concepts serve as a unifying force which makes vocational agricultural education a singular program in the educational system of the nation.

1. Vocational agricultural education programs are developed and conducted as a part of educational systems and are in harmony with a total philosophy of education for the individual and the society.

2. The changes within the agricultural sector of our technological society require that the major effort of vocational agricultural education focus upon preparing individuals for work and for entrance into the work force or entrepreneurship.

3. Vocational agricultural education programs relate to the productivity of people in terms of competencies in agricultural occupations, attitudes towards the occupations, and a willingness to produce efficiency.

4. When vocational agricultural education programs are established, the opportunities within society and the needs of society will be considered, as well as the interests and competencies of the individual.

5. The quantity of vocational agricultural education programs will be in keeping with employment patterns at the local, state, and national levels, in that order.

6. Vocational agricultural education, rather than being classified as a discipline within the educational system, is a unique and identifiable program which combines the skills and technical content of various disciplines with the practical requirements of the world of work to prepare a person to succeed technically and socially.

7. The vocational agricultural education program is unique in its requirements for community resource utilization, facilities and equipment needs for instruction, curriculums, instructor qualifications, and student goals.

8. To assure quality, vocational agricultural education programs are responsive to the needs of the individual for job-entry skills, and the compatible skills of communication, citizenship and leadership, decision-making, attitude to learn, and personal and occupational responsibility.

9. Vocational agricultural education programs possess a time commitment of sufficient length and intensity to provide instruction important to the successful entrance of the student into and advancement within the chosen occupations or entrepreneurship.

10. Vocational agricultural education programs are developed and conducted with individuals representing business and industry in the occupational area in which the program is being offered serving in an advisory capacity.

11. Vocational agricultural education is a part of the career development continuum, which includes:

a. Education for choice of an agricultural occupation through career motivation, career orientation, and career exploration,
b. Education for entrepreneurship or employment--vocational agricultural education.
c. Education for upgrading and retraining--vocational agricultural education.

12. Vocational agricultural education starts via defining occupational objectives, providing preparation for a job in agriculture and ending with the individuals' successfully entering entrepreneurship or jobs in agriculture, and such individuals will have capacities to continue to learn and transfer personal and occupational skills to meet the changing job requirements of the agricultural sector of a technological society.

13. Vocational agricultural education programs are available for youth at both the high school and post-high school levels, and for adults throughout their working life.

Other Developments

Contemporary developments in the areas of defining terminology and defining competencies in agricultural/agribusiness occupations are also of importance to teacher education in agriculture.

Defining Terminology. Agreement on terminology used in the profession is one criterion of maturation. The first definitions of terminology report was published by the American Vocational Association in 1954 (1.46). This publication included all vocational and practical arts areas. An updated terminology publication was attempted in the early 1970's, but consensus by all areas of vocational education was not achieved. This indicated the growing complexity and differences between the various areas of vocational education.

In 1976, the Publications Committee of the Agricultural Education Division of the American Vocational Association focused on the development of a terminology publication for agricultural education. The new terminology publication (1.47) developed is currently in press at the American Vocational Association.

Defining Competencies in Agricultural/Agribusiness Occupations. Teacher education must adequately prepare teachers to conduct local programs that reflect employment in farming and off-farm business and industry. Thus, information on the nature of the competencies needed in these occupations is significant to the preparation of teachers.

The first national effort, which began in 1975, was to identify census job titles that required agricultural competencies. This cooperative effort between the federal government and the agricultural education profession continued in 1976-78, with the objective of identifying and validating the competencies needed for entry and advancement in major agricultural/agribusiness occupations. The 1978 report provided an analysis of 57 production agriculture and 139 agribusiness occupations (1.48). The effort was conducted by 40 colleges and universities across the United States.

Toward Innovation

As new members are taken into a profession, they are traditionally admonished of the challenges that lie ahead. Whether the challenges of the future are more significant than those of the past is doubtful; but that they will possess their own unique characteristics is certain. These challenges bring a continuing series of developmental tasks and a satisfying confrontation with new ideas which are stimulators for further advancement of the profession, including its newest members.

Need for Innovation

In teacher education in agriculture, a program, supported by many tested principles, has evolved. New teacher educators begin with an existing program, as a child begins in the culture to which he/she is born. It includes practices, principles, and philosophies which are judged as representing the best yet attained. At the same time, leaders of the profession are seeking to advance knowledge and develop innovations in programs to meet their unique challenges.

A period of rapid growth is essentially dynamic. It is expected to be characterized by change and innovations. However, periods of declining school enrollments, such as in the current one, also result in considerable change. The specific innovations which will be developed and tested are not known. Nevertheless, it is reasonably certain that (1) some ideas transcend periods of change, thus creating an element of stability essential to orderly development and (2) the process of change is itself a subject of investigation, hence amenable to some regulation by individuals and groups.

Conditions Favoring Innovations

Spalding suggests that more is known about how to produce change than about what is good teacher education; thus, teacher educators in agriculture should develop proposals for change, with a regard for procedures to be used in carrying them out (1.49).

Studies of innovations as related to organization are relatively new. Some of the beliefs of and hypotheses on the innovating organization are reported in Transaction (1.50). These reports and Lippit's Dynamics of Planned Change are the principal bases for the following suggestions which are generally applicable.

1. Provide frequent informal opportunities in which staff members are confronted with problems. This would involve college coworkers,

supervisors, teachers, and others, but not necessarily at the same time.

2. Enlist voluntary association of two to five persons in studying a problem and developing a proposal. There may be advantages in having two or more groups competing to solve the same problem.

3. Provide the resources required; without them do not expect too much.

4. Arrange for some periodic variation of work and associations. (Intercollege or college-school contacts benefit all.)

5. Schedule periodic reviews or reports of progress.

The Focus of Innovation

Teacher educators in agriculture participate in the processes of change and are influenced by them, both in terms of professional progress and personal reactions. They become members in an institution with an established organization involving administrative officers, committees, and faculty bodies with certain governing powers. In addition, there is an on-going system of education, including agriculture, in the public schools with which teacher educators are involved. Hence, teacher educators concerned with change and innovations operate within a relatively complex network which includes many different interests and institutions. They will resist some changes or innovations and favor others. In terms of the viewpoint expressed in this chapter, they are obligated to analyze proposals carefully and to test, if possible, their impact on teacher education.

Change in teacher education is manifested in the behavior of the teachers prepared. Changes which make no difference in behavior are probably inconsequential. Those which lead to predetermined improvement in behavior are presumed to be wise. Those which by circumstances lead to chance improvement in behavior are fortuitous. Work should be directed to bring about predetermined improvements. This implies a continuing task of defining desired behavior (the model or proper role) and developing a program of teacher education in terms of maximizing its influence on students in these directions. It further implies the existence of an organization working toward the ideal of maximum effectiveness.

Summary

Teacher education in agriculture, a professional subdiscipline, is an innovation of this century. Its inception followed a growing demand for teachers of agriculture in public schools, and it achieved regularized status after the passage of the Smith-Hughes Act. It developed, generally, as a part of each state's land-grant college, with assistance from both state and federal educational agencies. Major efforts were devoted to preservice and inservice education of teachers of vocational agriculture. By 1966, it was established in 76 colleges or universities, of which 63 were land-grant institutions. The number of teacher educators employed at these institutions was about 260, but almost half of the institutions had no more than two positions each in this category. By 1979, despite a declining school age population, teacher education was in 93 institutions, and teacher education staffs had increased to 377, a 45 percent increase.

Although the profession was organized in various departments and operated under diverse conditions, a gradual unification and a growing identity with professional education took place. It was initially directly involved in providing leadership for the development of local programs of vocational agriculture and, influenced by forces of the times, gave emphasis to efficiency of agricultural production. Change to reflect employment in off-farm agribusiness occupations has now occurred.

It can be generalized that the period of 1967-1981 was one of maturation for the agricultural education profession. This is evidenced by national efforts and progress toward consensus and application of philosophy, terminology, standards, and competency identification.

Innovation in teacher education will continue to be needed to prepare teachers to conduct programs that reflect the rapidly changing nature of agriculture/agribusiness.

CHAPTER 2

The Need for Teacher Education in Agriculture

Clifford L. Nelson
UNIVERSITY OF MARYLAND

Writing the justification for one's own profession should be the easiest task one faces as a professional. Often in government and business the first assignment given to a new employee is to write a "job description." This chapter will be in the form of a professional position description, with operational definitions of the needs that are to be met.

Background

College faculty members in agricultural education, who are working as teacher educators in agriculture, face challenges not addressed by state supervisors or teachers in the field (2.1). In most states, teacher educators are full-time employees of their respective universities or state colleges. Direction and funding from the state department of education is usually through support of the university position(s) by formula or project rather than as employer. Therefore, being only responsible in part to the state department of education, the teacher educator remains primarily a university employer. Thus, faculty members are faced with the need to perform those tasks that lead to promotion and tenure within the university. The publish-or-perish syndrome affects teacher educators significantly. Many of the activities most popular with state supervisors and teachers in the field are not rewarded at the university. Consequently, the teacher educator, who is usually a former teacher and/or supervisor, is often faced with a conflict of interest between known needs in the field. The conflict as to what the role of the teacher educator should be also exists between university administrators and state supervisors. Furthermore, teachers in the field have a different perception of the role of teacher educators than do state supervisors, university officials, and teacher educators themselves. It is not the purpose of this discussion to resolve these conflicts, but rather to set the stage for the development of a justification for teacher education in agriculture within the reality of this milieu.

The Need for Teacher Education in Preservice Education

Teacher education in general has followed several trends in recent years that have made the justification of traditional preservice teacher education more challenging. The trend of establishing departments of vocational education has taken away from agricultural education departments much of the autonomy in course content and curriculum content and activity in teacher education (2.2). More and more introductory courses in agricultural education have been retitled vocational education and have been expanded to include all vocational students. Teacher educators in agriculture are often teaching several courses for vocational students rather than instructing only agricultural education students.

Many colleges of education have developed student teaching centers where student teachers from more than one discipline are placed. Supervision at the centers is often carried out by university faculty members who have no agricultural education background. The agricultural teacher educator may not see an agricultural student teacher during his/her student teaching. The development of general methods courses in vocational education and education has in many instances diminished the role of the agricultural teacher educator to being one of advising students and offering one or two courses in agricultural education at the undergraduate level.

The need for undergraduate courses in agricultural education and teacher educators qualified to teach those courses is reflected in most state plans for vocational education. A typical state plan for vocational education, one that will be supported by vocational funds, describes the qualifications of a teacher educator. Most plans require three to five years of successful teaching experience before a professor can be designated as a teacher educator. Similar need is expressed in standards for agricultural education.

Cox and McCormick point out that the background and experience of prospective teachers has changed drastically in the last decade (2.3). When looking at undergraduate enrollments, most teacher educators see: ". . . former students of agriculture are clearly in the minority. Most have not lived on a farm or worked in other areas of agricultural industry." This further supports requiring specialized experience and preparation for teacher educators in agriculture.

Vocational education in agriculture is a broad and diverse field both in the subject matter it covers and in the activities it encompasses. The sciences of botany, zoology, chemistry, entomology, etc., are central to the study of agriculture, as are the applied sciences of animal science, plant science, agricultural engineering, agricultural economics, forestry, and natural resources. The fields establish the broad parameters of the subject matter taught by teachers of vocational agriculture.

Not only do secondary vocational agriculture programs have the traditional classroom activities of other subjects, but they also include activities in the in-school laboratory and activities in the agricultural mechanics laboratory (shop), with all its metal and woodshop skills in farm power machinery, construction, and electrification. In addition, many school programs include activities in laboratories such as greenhouses, nurseries, school farms, tree farms, and animal care facilities. Out-of-school supervised occupational experience programs on home farms and other farms and at agricultural businesses and public sites are also included in the agriculture teacher's responsibilities.

The agriculture teacher also advises the Future Farmers of America (FFA), which is one of the most active youth organizations found in secondary schools. The range of curriculum-related contests and activities is numerous as are the leadership development activities of this organization. Although many FFA activities are held during regular school hours because FFA is intracurricular, many activities are held after school hours. This includes contests and the practice and preparation for each.

The agriculture teacher is a businessperson. Inventories of school and FFA equipment and supplies are maintained. A school agribusiness is often operated with the produce of a greenhouse or school farm. Similar activities for the FFA chapter are often carried out in conjunction with the laboratory produce. One Maryland FFA chapter had over $40,000 in transactions in one year (2.4). This is not an atypical activity for an active chapter.

The multitude of activities with which a local vocational agriculture teacher is involved demonstrates the complexity of the position. When it is compared to other teachers' job descriptions, the business, advisory, and subject matter requirements for the teacher of agriculture far surpass the responsibilities placed on other teachers (2.5). Although some teachers have some of the roles expected of agriculture teachers, few, if any, have all the broad range of responsibilities facing agriculture teachers.

Therefore, it seems logical that a teacher educator in agriculture should have experienced some of the many requirements faced by teachers in the field. Teaching how to advise the FFA without having experienced the activity would be sterile and academic, just as would teaching dehorning a calf without ever having performed the task.

Vocational education in agriculture is based on community needs. English, history, mathematics, science, etc., are taught much the same across the state or the nation. However, vocational agriculture concentrates its curriculum on the agriculture and the agribusinesses found in the school's service area. Community orientation makes curriculum development a unique and challenging task, when compared to the curriculum development in other fields. It is not uncommon to see widely varied agricultural curricula at secondary schools in the same county, depending on the school's service area. Teaching this unusual curriculum development requires teacher education techniques that are quite different from those that might be employed in other educational disciplines.

Community demands on teacher time are greater for the vocational agriculture teacher than for most teachers. Active farm organizations, service clubs, community fairs and shows, as well as garden and 4-H clubs, demand and require commitment from the vocational agriculture teacher.

The preceding brief description of the vocational agriculture (vo-ag) teachers' tasks leads to an operational definition of what a teacher educator in agriculture must be able to do (2.6). If the same logic is used for teacher educators as for vocational agriculture teachers, it then follows that a teacher educator must be an experienced vocational agriculture teacher who has performed the large majority of tasks that he/she is expected to prepare preservice teachers to perform (2.7). It also follows that a teacher educator whose specialty is outside agriculture education would have difficulty in being a successful teacher educator in agriculture (2.8).

The Need for Teacher Education in Inservice Education

No undergraduate curriculum and/or combination of skilled teacher educators has yet to prepare the preservice teacher of agriculture fully for the role he/she will perform (2.9). For this reason, inservice education typically concentrates first on beginning teachers. The "first-year" teacher courses and special inservice meetings, as well as personal follow-up and visitation of first-year teachers, have long been a part of inservice teacher education.

Inservice education is also needed for experienced teachers in both technical and professional education subject matter (2.10). Technical agriculture subject matter is constantly changing. Subject matter taught as an approved practice in many freshman college courses may be outdated by research before a student graduates. Textbooks are dated before their publication. University and industrial research contribute to continuing change in technical agriculture information.

Many teacher educators are located at the source of much of the new technical education--the state land-grant colleges. In addition, the land-grant colleges serve as clearinghouses of research information from other sources through the cooperative extension service.

Teacher educators are usually located in ideal sites for the coordination of the dissemination of updated technical information. Most state extension specialists are located at land-grant institutions, as are the technical agriculture department instructors who do undergraduate and graduate teaching. Dissemination of technical information through an instructional materials service is an important part of inservice education provided by most teacher education in agriculture programs.

Access to technical departments in most universities assures the opportunity to seek qualified teaching faculty to offer inservice and graduate classes to teachers. In some cases, access to academic expertise and time to serve teachers directly has been limited. Therefore, certain agricultural education departments have employed their own subject matter specialists to serve as liaison between technical agriculture departments and the teachers in the field. These subject matter experts usually teach short courses, develop and conduct workshops, and create and produce teaching materials for teachers.

In addition to coordinating delivery of technical agriculture subject matter, teacher educators also have the responsibility for inservice in the area of professional education inservice in agriculture. This need is demonstrated by the changes in vocational education in agriculture since the 1963 Vocational Education Act and the 1968 and 1976 amendments to the 1963 Act. In addition, the effects of the Education for All Handicapped Children Act of 1975, P.L. 94-142, which mandates free and appropriate education for all children in the least restrictive environment, has made, and will continue to make, significant changes in the operation of vocational education in agriculture. Coupled with the changes mandated by federal laws, there has been significant change in state legislation affecting the daily operation of education. Some examples are state-required reading courses, statewide-required competency-based curricula, and the "return-to-basics" legislation that is found in many states.

Pre-1963 FFA members and advisors would recognize few of the many programs of the 1980's. Prior to 1963, the FFA was limited to production agriculture and included only male members. Not many changes were introduced. However, since then, the FFA has increased the number of programs, included females in its memberships, and expanded its awards to reflect vocational education in agriculture more accurately. These changes have greatly increased the need for teacher education in the FFA. Special texts and teaching materials have been prepared for preservice and inservice education of teachers. A special series of nationwide and regional meetings was conducted in 1979 just to introduce new FFA programs and materials.

Teacher educators, usually with the cooperation of the state supervisory staff, conduct studies of teacher inservice needs. Several areas, in addition to the traditional technical agriculture and the FFA-related needs, have regularly been included. Examples include state record books, pesticide certification, occupational safety and health requirements, tort liability of teachers, and metrification in agriculture. Depending upon the complexity of the subject, the length of inservice education devoted to a given topic might range from a 30-minute presentation at an agriculture teachers meeting to a four-credit graduate course conducted over the course of a semester or a school year.

Teacher educators in agriculture have a unique background of general education in agriculture, secondary teaching experience in agriculture, and advanced degrees in vocational education or agricultural education. They are therefore best qualified to determine inservice educational needs, to coordinate subject matter offerings, and to conduct direct educational activities for beginning and experienced teachers of agriculture.

The Need for Teacher Education for the Extension of Knowledge

Teacher educators in agriculture are typically not cooperative extension service faculty members. However, several states have seen the need for the cooperative extension service to offer services to teachers formally through state extension specialists, just as extension has served other client groups. Supported or not by cooperative extension, teacher educators have a major responsibility in the extension of knowledge about teaching. This extension of knowledge is via inservice and preservice education as described and through writing and personal contacts. Teacher educators located at universities and colleges have available the latest research results and journals. Few teachers in the field maintain subscriptions to Summaries of Studies in Agricultural Education or to the major research journals in education, vocational education, and teacher education. Many of the larger school systems now maintain microfiche collections of the Educational Resources Information Center (ERIC) and the Current Index to Journals of Education (CIJE), as well as selected educational journals. However, unless the classroom teacher is actively involved in graduate work, this availability is not typically known or used. State newsletters are regularly shared among teacher education faculties. Many articles of interest and possible application are published in the newsletters.

The professional obligation of the teacher educator is to make the largest research results and successful innovations from other states available to students and teachers through their preservice and inservice activities (2.11). The teacher educators' regular visits to local schools and interaction with teachers provide opportunities to assess local and individual needs of teachers and to offer new techniques, methodologies, and programs that have been suggested by research or applications elsewhere.

The Need for Teacher Education in Research

Involvement of teacher educators in research is mandated by two major factors. The first is that teacher educators are typically employed as professionals in institutions that determine promotion, salary increases, and prestige mainly on faculty research and writing. The second factor, and for many the more important, is the need for (1) continuous research to improve the quality of agricultural education, and (2) improvement in teacher educators' professional preparation and experience in conducting research.

Just as the local teachers of agriculture must take their turn at hall duty, bus supervision, and homeroom, teacher educators must perform the duties expected by the institutions that employ them. Universities and colleges are seldom administered by vocational educators. Most administrators have risen through the ranks of academic professorship and administration via their research and the reputation created therefrom. Professors in agricultural education are usually evaluated on the same criteria as other faculty members, and thus, they must conduct research and publish in order to survive. However, in evaluating for promotion and advancement, some universities and colleges do recognize service contributions and publishing in newsletters and non-research-oriented journals by teacher educators.

Viable education programs are characterized by their ability to change with the changing world (2.12). Competently conducted research addressing relevant questions should be the basis for recommending changes in programs. Change will take place in the schools. Whether the change is rational will depend upon the research information available to decision makers (2.13). The agricultural teacher educator is best prepared to conduct this research because of experience in the field both as a teacher and as a teacher educator.

Teacher educators in agriculture, along with state supervisory staff persons, are most knowledgeable about local school conditions and needs. If teacher educators in agriculture do not conduct the needed research, researchers in other areas will select the research questions, determine the data to collect, and perhaps interpret the data outside the context of the real needs of agriculture.

Research on questions concerning agricultural education should be encouraged. Research competently and objectively conducted by "outside" individuals and institutions can enhance agricultural education. Objectivity can only be assured when the findings of agricultural education research support results generated by outside research.

Research in agricultural education has shown that Department of Labor (DOL) statistics on the need for graduates prepared in agriculture are often inaccurate. Many studies conducted within school districts, regions, and states have demonstrated the inadequacy of DOL statistics. Each year new DOL human resource data which exclude many occupations requiring agricultural competencies are published. Continuing research is needed to identify agricultural occupations that are excluded by the human resource reports.

Accurate employment demand data is only one example of research needed to maintain current programs. Continuous research in new and developing programs as well as in established programs is also needed. Research conducted by teacher educators will likely be more field based and applicable than research conducted by others. Questions that are of most importance to agricultural education are more likely to be asked by teacher educators. Research results are more quickly communicated and put into use when teacher educators who conduct the research are those involved in inservice, preservice, and extension activities.

The Need for Teacher Education to Represent Vocational Education in Agriculture

Agricultural educators are constrained by their professional positions in representing the viewpoint of their profession as a whole. Local vocational agriculture teachers are limited by the policies of their school districts, and the state supervisory staff must operate within the guidelines of the state department of education. Teacher educators in agriculture must operate within university regulations, and the staff persons in agricultural education at the U.S. Department of Education (USDE) must follow the policy of their unit. Vocational education in agriculture has faced many outside pressures that have threatened local, state, and national programs. In many cases, the individuals involved have not been able to offer leadership to counter external pressures because of their positions.

Reductions in supervisory staff at the state and national levels typically cannot be countered by state department of education or USDE employees. The leadership to create the need for these positions should come from local teachers and teacher educators. Since university and local school district policies usually do not preclude involvement in establishing the need for supervisors, these segments of agricultural education actively have been able to develop rationale and to generate support to continue professional agricultural supervision at the state and national levels.

Teacher educators can often offer more effective leadership at the local school district level than supervisors or other teachers might be able to provide. The prestige of the university, as well as the possibility of using current research, can enhance the teacher educators' influence.

Professionals--teachers, supervisors, and teacher educators--in agricultural education ideally operate in mutually supportive roles. They must be ready to represent their profession and to support the other segments. Teacher educators, because of their academic position, their experience at local and statewide levels, and their knowledge of research, have the responsibility and opportunity both to prepare quality beginning teachers and to assist the local teacher and state supervisor in developing good programs. Herein is the need for teacher education in agriculture.

CHAPTER 3

Programs of Teacher Education in Agriculture

Floyd G. McCormick
THE UNIVERSITY OF ARIZONA

Roland L. Peterson
UNIVERSITY OF MINNESOTA

"Teaching is the mother of all professions." This profound statement has far-reaching implications for the progress and survival of society. Of more consequence, it has strong implications for the preparation of individuals entering the teaching profession.

The quality of any instructional endeavor, whether formal or informal, is directly related to the competence of the individual teacher who is planning, directing, and evaluating that endeavor. Education at any level is only as effective as the teacher. Undergirding this basic tenet is the quality of the preparation program designed and utilized to prepare prospective educators. Thus, the quality of instructional programs at any level is, to a very large extent, influenced by the quality of the teacher education program preparing teachers to provide the instruction.

This chapter addresses the development of teacher education in agriculture programs in an attempt to identify the evolving functions and responsibilities existing in most of today's programs. Various program approaches to teacher education in agriculture are discussed to illustrate the ways institutions of higher education are designing and implementing education programs to prepare the teachers necessary to perpetuate the profession. The utilization of guidelines for teacher education in agriculture programs, along with the identification, legitimization, and implementation of program standards, is addressed in the light of program design. Other concerns affecting programs of teacher education in agriculture are discussed briefly at the end of the chapter.

Development of Teacher Education

In the broadest sense, the training of individuals to take the place of those leaving the "trade" is probably as old as humankind itself. In the same context, the preparation of teachers by informal means is likewise as old. As O'Kelley states, the first teacher educators in agriculture probably were those who prepared the teachers of the early settlers of the colonies. He further points out there is unfortunately little or no record of the techniques and procedures utilized in these pioneer teacher education programs (3.1). It can be assumed, however, that no one pattern for such programs emerged; nor is there a record of the establishment of continued organized programs for teacher education in agriculture.

The Formative Years

One of the publications detailing the historical background of the development of agricultural education in each state was compiled by Stimson and Lathrop in 1942 (3.2). The Massachusetts Agricultural College was probably the first college in the United States to establish a department of agricultural education. The establishment of that department occurred in 1907, and establishment in other institutions of higher education followed soon thereafter. Departments of agricultural education for the purpose of training teachers of agriculture were established at Michigan Agricultural College in 1908, Iowa State College in 1911, and The University of Minnesota in 1912. Following the passage of the Smith-Hughes Act in 1917, a major impetus was generated to establish additional agricultural teacher-training programs. By the year 1922, 48 land-grant institutions had organized departments of teacher education in agriculture. In 1978, there were 85 institutions of higher education offering teacher education in agriculture programs.

A review of the history of teacher education in agriculture development shows a lack of uniformity existing in these programs. This diversity in programs has been due, in part, to the following:

Administration (Location) -- Approximately one-half of the departments of agricultural education are administrated as a part of the college of agriculture rather than the college of education or some other unit preparing teachers.

Institutional Philosophy -- A wide variation in institutional philosophy and organization prevails in higher education circles in the various states. Federal control with regard to agricultural teacher education has been minimal. Every attempt has been made to safeguard "local direction and control" of teacher education programs. As a result, teacher education in agriculture programs reflect the philosophical base of the institution in which they are located.

Lack of Guidelines -- The identification of acceptable principles, guidelines, and program standards upon which to build a national program has been lacking. However, formulation of an agreed-upon philosophical base for specifying the breadth, the scope, and the intent of programs has received major attention in recent years (3.3).

It should be pointed out that only in the last 20 years have efforts on the national level been undertaken to develop, in published form, guidelines and standards which could be used by teacher education in agriculture programs to provide program direction as well as criteria for evaluating the relative program effectiveness. These efforts are discussed later in this chapter and in Chapter 13.

Expansion of Purpose

Study of the development of agricultural teacher education programs shows that the functions and responsibilities of programs have expanded greatly over the years. Programs of teacher education in agriculture initially addressed only the "teaching" function; that is, there was major emphasis on teaching prospective teachers those pedagogical skills essential for them to plan, conduct, and evaluate vocational agriculture programs at the local level.

Although the teaching function is still the primary focus of most teacher education in agriculture programs today, the purpose has expanded rapidly in the past two decades, especially since the passage of the Vocational Education Act in 1963. Teacher education programs have been expanded to assume functions such as research; inservice education; beginning teacher supervision; curriculum development, dissemination, and evaluation; college teaching improvement; extension education; and international education. This shift over time from primarily one major function (teaching) to a multi-functional role has created a demand for additional fiscal and human resources to carry out the broadened mission of teacher education in agriculture. From this demand has evolved a critical need to plan strategies relative to the allocation of limited resources to those functions which (1) are within the domain of the teacher education program and (2) which provide the greatest benefit to the agricultural education profession. The trend towards broader-based program functions began to evolve rapidly once the profession took official action regarding the adoption of a set of guiding principles which, in turn, served as a vehicle to widen the overall program objectives and resultant functions.

Although the functions served by most agricultural teacher education programs have expanded, not all programs are involved in all functions. The size of the teacher education staff and fiscal resources available dictate, to a large extent, the degree of involvement and participation in inservice education, curriculum development, extension education, and international education. However, a majority of agricultural teacher education programs in the United States address the broad missions of instruction, research, and professional leadership, development, and service. This is as would be expected since a majority of the teacher education programs in agriculture are located in land-grant institutions of higher education. Major emphasis on resident instruction, research, and service are inherent in these institutions.

Program Principles and Guidelines

As previously described, a great diversity existed during the early development of programs of teacher education in agriculture across the United

States. This diversity has been due, in part, to the lack of written and accepted guidelines to provide direction for maintaining, upgrading and updating, and initiating agricultural teacher education programs.

Early Efforts to Identify Guiding Principles

O'Kelley states that apparently during the first 25 or 30 years after the passage of the Smith-Hughes Act, agricultural teacher educators were too deeply engrossed in the problems of program development to concern themselves with academic details such as keeping a record of the principles which undergirded their decisions and programs (3.4). They were concerned primarily with the continuing task of training teachers for jobs in a constantly evolving program of vocational education in agriculture.

In 1953, the Teacher Education Committee of the Agricultural Education Division of the American Vocational Association became interested in the development of guiding principles for training teachers of vocational agriculture. Dr. V. G. Martin, Chairman of the Teacher Education Committee, prepared a tentative draft of principles which were first considered by his committee at the AVA convention in 1955. These principles were reviewed the following year by the teacher education section of the Agricultural Education Division at the 1956 AVA convention held in St. Louis. In 1958, the American Vocational Association published Guiding Principles for Institutions Training Teachers of Vocational Agriculture (3.5). This was the culmination of work performed by the Teacher Education Committee over a period of several years. The following principles were proposed in the report.

Teacher-training institutions should:

1. Be responsible for providing adequate facilities and staffs for training teachers of vocational agriculture.

2. Make provisions for recruiting, training, and placing prospective teachers of vocational agriculture.

3. Make provisions for inservice training of teachers of vocational agriculture.

4. Make provisions for conducting research essential to the program of vocational agriculture.

5. Be responsible for processing and making available teaching aids needed by teachers of vocational agriculture.

6. Cooperate with local, state, and national groups and individuals concerned with the welfare of education and agriculture.

7. Be responsible for continuously appraising changes in agriculture and for making curriculum adoptions to meet varying conditions.

These seven principles reflect the areas of responsibilities to be assumed by institutions offering training programs for teachers of vocational agriculture, and they indicate a broadening role for teacher education in agriculture. It is interesting to note that there was no mention of graduate education, extension education, international education, or follow-up and supervision of graduates in the 1958 Guiding Principles. However, these principles do show that the leadership in agricultural teacher education was considering a broadened base for teacher education as early as 1958.

Preservice Guidelines

The Teacher Education Committee of the American Vocational Association was still not satisfied that the specific responsibilities of teacher-training institutions relative to the undergraduate or preservice training of teachers had been clearly identified, so it continued with its deliberations at the 1961 AVA convention in Kansas City. This committee presented a report entitled Guiding Principles for Pre-service Training of Teachers of Vocational Agriculture (3.6). The report was approved and contained the following statements:

> The undergraduate preparation must be based on the responsibilities teachers of vocational agriculture are expected to assume in their work in the school and in the community. Such responsibilities encompass educational programs for adults engaged in or concerned with agriculture; for young people interested in becoming established in farming or other agricultural occupations; and for high school youths including F.F.A. and N.F.A.
>
> I. THE CURRICULUM--GENERAL CONSIDERATIONS
>
> The Pre-service Teacher Training Curriculum
>
> A. Provides opportunities and experiences to develop personal, social, and professional qualities exemplified by superior teachers of vocational agriculture.
>
> B. Places emphasis on basic scientific concepts, principles, and relationships.
>
> C. Provides functional and practical education.
>
> D. Provides for integration of theory and practice.
>
> E. Allows flexibility and recognizes needs and interests of individual student.
>
> F. Provides reasonable depth of specialization.

II. GENERAL EDUCATION

The general education of the teacher is at least as comprehensive as that of college graduates who enter other fields of public secondary school teaching.

The general education of the teacher will develop the ability to:

A. Communicate ideas clearly and effectively.

B. Understand and apply democratic concepts.

C. Think in terms of local, state, national, and international problems.

D. Participate effectively in civic affairs.

E. Identify and develop a philosophy of living.

F. Appreciate the cultural and aesthetic aspects of our society.

G. Appreciate the scientific and experimental approach.

H. Work effectively with people.

III. TECHNICAL AGRICULTURE

A. The curriculum gives students an opportunity to acquire such technical knowledge and skill in plant and soil science, animal science, agricultural economics, and agricultural engineering as are necessary to conduct a superior high school program of vocational agriculture.

B. The curriculum includes courses in technical agriculture that provide opportunities for the student to:

1. Interpret and use scientific information.

2. Supplement classroom theory with practice.

3. Develop enough skills to meet the needs of a typical beginning teacher and knowledge of how to acquire additional skills as needed.

IV. PROFESSIONAL EDUCATION

The curriculum provides professional education courses that:

A. Include an understanding of the process of human growth and development; the mental, emotional, and physical behavior of learners; and the psychology of learning, adjustment, motivation, and personality.

B. Include an understanding of the purpose, structure, administration, and operation of the school system including the range of education programs and curriculum patterns. (The teacher needs to know how his own job relates to the whole school program and the personnel involved in the educational enterprise. Such knowledge is needed in order that the teacher may more adequately fulfill his role as a teacher and take his place as a responsible member of the profession.)

C. Include training in special methods and techniques in program planning and conducting a complete program of vocational agriculture. (This includes studying, evaluating, and utilizing community resources, organizing and evaluating subject matter for teaching purposes, and planning for effective teaching. Such a program encompasses education for adults engaged in or concerned with agriculture; for young people interested in becoming established in farming or other agricultural occupations; and for high school youth including F.F.A. and N.F.A.)

D. Include supervised experiences that are planned for induction into the job. (This should include observation, participation, student teaching, and competence in methods for teaching and in the use of instructional resources.)

The terms "teacher training" and "NFA" are used in the preceding list of guiding principles. The term "teacher training" is seldom used in today's literature, and the New Farmers of America (NFA) was integrated into the national Future Farmers of America (FFA) organization in 1965.

The historical record shows that it took almost 45 years for the agricultural teacher education profession to adopt a set of guiding principles. These efforts generated a written philosophical base for teacher education in agriculture programs through the identification of specific guidelines.

A Consensus on Principles

In the first edition of this book, O'Kelley presented what he felt reflected agreement among agricultural teacher educators on general principles at that time. These tenets were (3.7):

1. An effective agricultural teacher education program provides for both the preservice and the inservice training of teachers.

2. Not only is the effectiveness of the agricultural teacher education program grounded in sound research findings, but the program itself is also concerned with the continued production of such needed research.

3. The use of high quality instructional materials, which can be efficiently prepared by teacher educators, can greatly enhance the effectiveness of teachers in service. Teacher involvement

in such preparation can also result in an improved teacher education program.

4. Trainee recruitment and follow-up of graduates contribute to the overall effectiveness of the preservice agricultural teacher education program.

5. The overall competency of an agriculture teacher is directly related to the quality of his/her training and development in the areas of technical agriculture, general education, and professional education.

6. The best measurement of the effectiveness of an agricultural teacher education program may be the ability of its graduates to perform proficiently in the type of situations for which they are trained.

7. The provision of opportunities for teachers in training to gain supervised participating experience in situations typical of those confronting teachers in service increases the effectiveness of the agricultural teacher education program.

8. The fully effective agricultural teacher education program, in addition to its teacher training activities, provides leadership training for those who are responsible for planning, administering, and supervising the overall agricultural education program.

9. Agricultural teacher education programs function most effectively in a climate where teacher educators and those responsible for the conduct of public school programs, including instruction in agriculture, work harmoniously and cooperatively in the planning and the conducting of the total program.

10. Agricultural teacher educators who exert professional influence through accepted institutional channels toward the end that the kind and the quality of technical studies available to preservice and inservice agriculture teachers are consistent with the needs of such teachers are acting in the best interests of the agricultural teacher education program.

Contemporary Efforts to Identify Program Guidelines and Standards

In the early 1970's, the American Association of Teacher Educators in Agriculture and the Agricultural Education Division of the American Vocational Association commissioned an ad hoc committee to identify recommended standards for teacher education in agriculture. The desired outcome was to synthesize a set of guidelines which would set down the parameters by which programs of teacher education in agriculture might be improved. The goal was to bring about a more effective delivery system and, at the same time, to provide criteria for carrying on a systematic program of evaluation. In addition, it was an attempt by the profession to develop a philosophical base for teacher education in agriculture programs through the identification of specific guidelines. The guidelines could serve as a means of reducing diversity and promoting a greater degree of unity in agricultural teacher education.

The AATEA guidelines addressed the following areas and sections (3.8).

AREA I. The Program of Instruction

1) General Studies
2) Content for Teaching Specialty (Technical)
3) Professional Studies

AREA II. Faculty in Teacher Education

1) Competence of Faculty
2) Utilization of Faculty
3) Faculty Involvement in Schools
4) Conditions for Faculty Service
5) Part-time Faculty

AREA III. Students in Teacher Education

1) Admission to Teacher Education Program
2) Screening on the Basis of Academic Achievement
3) Screening on the Basis of Personality Characteristics
4) Student Personnel Services
5) Student Involvement

AREA IV. Resources and Facilities for Teacher Education

1) Library
2) Materials and Instructional Media Center
3) Facilities
4) Clerical and Supporting

AREA V. Evaluation Program Review and Planning

1) Evaluation of Graduates
2) Use of Evaluation Results to Improve Basic Programs
3) Long-range Planning

In the process of program planning, it is essential to recognize the important role standards play in the successful implementation and eventual delivery of quality teacher education in agriculture programs. Most teacher educators believe that program standards should be set forth for all aspects of the program, whether it be for (1) maintaining existing programs, (2) upgrading existing programs, or (3) initiating new programs. There is an imperative need to identify those pertinent program standards that should be considered when planning the design and content of relevant teacher education programs.

Program standards are educational specifications describing the essential ingredients inherent in a quality education program. Such standards should encompass specifications dealing with the (1) instructional program, (2) instructional staff, (3) student enrollees, and (4) physical facilities.

Written as criteria which should be measurable and specific in nature, they should serve as a basis for program evaluation.

The need for a set of national criteria which could be used to upgrade existing programs and to develop new agricultural teacher education programs was recognized during a national seminar on agricultural education held in Denver, Colorado, in May 1971. Between 1971 and 1976, the decision was made to identify a set of program standards for agricultural education. The initial step was taken during a three-day seminar held in Kansas City, Missouri, during the spring of 1976. As an outcome of this seminar, standards were identified for each of the seven major taxonomy areas for both secondary and post-secondary programs and for teacher education, supervision, and administration (3.9).

Following the 1976 national seminar, the Department of Agricultural Education at Iowa State University was awarded an Educational Professions Development Act (EPDA) contract to validate the standards and to develop a dissemination and implementation plan. The standards identified and verified would serve as guidelines for the various states to utilize in legitimizing and eventually implementing specific state program standards (3.10).

Utilizing the list of standards specific to teacher education, the American Association of Teacher Educators in Agriculture commissioned an ad hoc committee, chaired by Oren of Mississippi State, on the legitimization of program standards and guidelines. The purpose of the committee was (1) to determine the relative degree of acceptance by teacher educators in agriculture of the national standards for teacher education programs in agriculture/agribusiness and the guidelines for teacher education; (2) to ascertain how these standards and guidelines were being implemented in teacher education programs; and (3) to make recommendations to the profession.

The final report of this committee was presented at the AVA convention in Atlantic City in December 1977. The following recommendations were proposed in the report (3.11):

1. The standards for teacher education in agriculture should serve as a benchmark to guide program planning and evaluation.

2. The standards should not be construed to be the only criteria for program evaluation.

3. The AATEA Board of Directors should commission a standing committee on national standards for teacher education in agriculture.

Many teacher education in agriculture programs are currently utilizing the validated standards as guidelines to improve, upgrade, and update their teacher preparation programs. The standards are also serving as valuable criteria for comprehensive program evaluations.

The dynamic nature of agricultural teacher education programs requires that periodic evaluations be conducted. Program standards serve a useful purpose in planning and evaluating processes, but only if these standards are current and if they are in line with contemporary programs. Therefore, it is imperative that teacher educators continue efforts directed toward providing up-to-date guidelines and standards for the improvement of teacher education in agriculture programs.

Program Mission

It is generally accepted that the overall mission of higher education is threefold: instruction, research, and service. More specifically, the general goals of higher education institutions are (1) to provide the opportunity for the acquisition of comprehensive education and usable skills; (2) to serve as a resource for the expansion of knowledge through research; and (3) to increase the opportunity to improve the quality of life by making available the services and resources of the institution to the many publics served. In essence, these general goals provide the vehicle to carry out the mission of institutions of higher education. Due to the nature of the land-grant colleges, the threefold mission is probably more pronounced in these institutions. Since teacher education in agriculture programs are located predominately in land-grant institutions, the major goals of higher education also become the foundation for defining the major program functions for agricultural teacher education.

Program Diversity

Diversity of college programs and curricular organizations is perhaps the major characteristic of higher education in the United States. Programs of teacher education in agriculture are no exception. A wide array of approaches are used to accomplish the functions of agricultural teacher education. Four-year training cycles, five-year cycles, competency-based programs, performance-based programs, block approaches, survival skills, dual majors, and internships are but a few of the diversified avenues utilized by agricultural teacher educators to prepare prospective agricultural educators. The range and extent of the approaches utilized to prepare prospective agricultural educators serves to illustrate the diversity existing in teacher education in agriculture. At the same time, there are basic similarities in programs. Agricultural teacher education programs have in common the three basic functions of higher education, that is, instruction, research, and service.

Thus, a paradox between similarity and diversity exists in teacher education programs. However, that diversity in program approaches contributes to the strength of the teacher education profession. In the final analysis, the approach used is not the major concern; rather it is the product of the program as exemplified by qualified teachers, relevant research endeavors, and services to teachers and programs.

A discussion of the predominate program approaches utilized in teacher education in agriculture will be presented later in the chapter.

Teacher Education Responsibilities

It is generally accepted that the program mission of teacher education in agriculture is clustered in terms of the general functions of instruction; research; and professional leadership, development, and service. Some of the specific responsibilities under these functions are:

A. Instruction

1. Undergraduate or Preservice Education

Involves recruiting, counseling, and placing students; planning students' programs of study; planning and teaching professional courses; supervising student teachers; and advising collegiate student organizations.

2. Graduate Education

Includes recruiting and selecting quality graduate students; planning graduate study programs; teaching graduate-level professional courses; planning and directing students' research programs; and serving on graduate committees.

3. Inservice Education (credit and non-credit)

Involves organizing and conducting short and intensive pedagogical and technical courses; planning long-range inservice education programs for practicing teachers; providing clinical supervision to new and experienced teachers; planning conferences and workshops; and developing individualized professional improvement programs.

4. Extension Education

Involves planning and teaching professional education courses for cooperative extension workers; developing individual professional improvement programs; coordinating inservice activities; and cooperating in planning and conducting seminars, workshops, and conferences.

5. International Education

Includes organizing and conducting specialized workshops and seminars; advising foreign students; cooperating with college departments to design interdisciplinary programs for students desiring to work in international agriculture; planning field trips and other intensive activities for visiting students and administrators from developing countries; and serving as advisors, teachers, and evaluators of programs in developing countries.

B. <u>Research</u>

1. Graduate Student

Includes planning and directing research studies of graduate students; preparing and editing theses and dissertations; and supervising the preparation and dissemination of research results.

2. Staff

Involves writing research proposals; initiating studies on problems in agricultural education; conducting research studies; preparing research reports; and publishing results of research findings.

3. Sponsored

Involves preparing requests for proposals; seeking outside funding resources; directing research; and publishing and disseminating research results.

4. Interdisciplinary

Includes identifying potential research projects; cooperating with other departments, colleges, and agencies in designing proposals and budgets; serving to coordinate research activities; and assisting in writing reports.

C. <u>Professional Leadership, Development, and Service</u>

1. University Relations

Involves serving on department, college, and university committees; supporting and contributing to collegiate activities; promoting the department, college, and university through public relations; and contacting high schools, community colleges, and agricultural businesses.

2. Field Supervision, Consultation, and Follow-up

Involves supervising new teachers in the profession; providing consultation and assistance to experienced teachers; directing the new workers program in extension education; and providing consultation services to high schools, community colleges, and technical schools for improvement of instructional programs in agriculture.

3. College Teaching Improvement and Consultation

Involves serving on committees designed to recommend ways to improve college teaching; developing evaluation procedures and approaches; participating in student-course evaluations; providing assistance and consultation to college instructors; and teaching courses on improvement of college teaching.

4. Program Planning and Evaluation

   Includes developing models, materials, and instruments for comprehensive program evaluations; promoting the development of long-range program plans; and conducting on-site visits to assist administrators and teachers relative to program planning and evaluation.

5. Curriculum and Instructional Resource Development and Evaluation

   Involves developing and maintaining instructional material centers; developing student and teacher references; conducting field-testing programs; and disseminating curriculum and instructional resource materials.

Teacher education in agriculture programs must address numerous objectives in order to contribute to the general mission inherent to an institution of higher education. Major emphasis should be placed upon long-range program planning as a means for achieving program objectives.

Program Goals and Objectives

The primary goal of teacher education programs is to develop students who will contribute to the improvement of society as active and informed citizens; as educational leaders; and as teachers of youth and adults (3.12). In the broadest sense, the preparation of agricultural educators should (3.13):

1. Provide educational opportunities and experiences that optimize the development of personal, social, and professional qualities.

2. Emphasize basic scientific concepts, principles, and relationships applicable to agriculture.

3. Provide functional, relevant education.

4. Allow for the integration of theory and practice.

5. Provide flexibility to meet the needs of individual students.

6. Provide a substantial measure of specialization.

A demand for versatile teachers has been created because of the breadth and depth of contemporary agriculture and the constantly changing instructional thrusts at the elementary, secondary, and post-secondary levels. The increasing demand for a variety of agricultural educators will place added responsibilities in the future on teacher education in agriculture. There is, and will be, a changing clientele to be served by the profession. Historically, programs of teacher education in agriculture primarily have focused upon preparing high school vocational agriculture teachers. At the present time, attention is being shifted to serve a wider clientele, ranging from formal education in general agriculture at the elementary and secondary levels to informal adult education (3.14).

Teacher education programs in the future will need to be designed to meet the professional and technical education needs for the following types of agricultural educators, as identified in Agricultural Education for the Seventies and Beyond (3.15).

1. Secondary teachers who will prepare youth for agribusiness careers.

2. Metropolitan secondary teachers who will prepare youth for employment requiring specialized competencies (knowledges, skills, and attitudes) in agriculture.

3. Post-secondary teachers who will prepare individuals for para-professional positions in agricultural industries.

4. Cooperative extension personnel who will serve a broad spectrum of youth and adult needs in agriculture.

5. Agricultural educators who will serve in international education endeavors.

6. Agricultural educators who will work in agribusinesses, including sales, promotion, and public relations.

7. Elementary and secondary school teachers who will teach general agriculture.

8. Agricultural educators who will conduct adult education programs.

9. Agricultural educators who possess specialized training for working with disabled and disadvantaged students.

At present, a few "flagship" teacher education programs have implemented extension education programs to prepare prospective cooperative extension workers. These undergraduate programs are competency-based approaches which have been integrated into the on-going preparation programs.

As has been the case for over 60 years, agricultural teacher educators in the future will assume responsibilities for new goals and objectives. The responsibilities assumed by teacher educators in agriculture are changing, as new activities continually are evolving. The demands placed upon teacher educators in agriculture in the future will be of increasing complexity. Leadership for identifying new directions for educational programs in agriculture rests heavily with teacher educators.

Program Approaches

In order to meet the professional needs, interests, and career objectives of future teachers, various program approaches have been utilized in the preparation of agricultural educators. "Program approach" as used in this chapter is synonymous with "type of program." The major intent of this section is to illustrate how teacher educators have designed delivery systems to prepare prospective agricultural educators. Program approaches as discussed here will be limited to professional education.

Most agricultural teacher education programs in this country can be grouped into three major categories, based upon the type of program approach used. These program types are the traditional, or course-based, approach, the competency-based approach, and the performance-based approach.

The ultimate intent of each program approach, especially at the preservice level, is to provide prospective agricultural educators with those knowledges, skills, and attitudes (competencies) required for initial success and advancement in the profession. Regardless of the approach utilized, each type of program is designed to expose students to those pedagogical skills required to plan, implement, and evaluate vocational programs in agricultural education.

Traditional, or Course-based, Approach

Out of necessity, agricultural education programs must be structured and organized around specific courses. Institutions of higher education traditionally have organized program curricula around a series of required and elective courses. In addition, most state certification agencies require completion of courses and credits in specific areas in order for prospective educators to receive certification to enter the teaching profession. Also, recertification in most states requires the completion of additional professional education courses. In the past, state certification agencies have approved teacher education programs on the basis of courses required for agricultural education majors. Hence, courses are a common basis for evaluating what is taught.

In the evolution of programs of teacher education in agriculture, the programs were designed based upon utilizing specific professional education courses. However, the course-based approach to teacher preparation focuses upon students completing the required number of courses in the proper course areas to meet degree and certification requirements. Because certain core education courses are required, there has been a tendency to have duplication of course content. Likewise, there is a tendency to design preservice programs around a series of unarticulated professional education courses in lieu of designing a professional preparation program that includes the professional competencies needed by beginning agriculture teachers. Similarly, there has been the tendency to minimize the application of pedagogical skills taught. Course-based approaches tend to become theoretical. Many traditional professional education courses have placed more emphasis on the theory of teaching rather than focusing upon specific teaching skills needed (3.16). The assumption is that application will take place at a later time during the supervised field experience program (student teaching). The integration of theory and practice is limited with this approach.

Competency-based Approach

One of the recent and perhaps far-reaching ideas in teacher education is competency-based teacher education (CBTE). Current literature regarding competency-based teacher education and performance-based teacher education (PBTE) tends to equate the two terms "performance" and "competency." Bender states that some people prefer to use synonymously the terms "competency-based teacher education" and "performance-based teacher education (3.17)." Emphasis here will be on illustrating how these approaches are, in fact, different.

Pedagogical content for both CBTE and PBTE programs centers on the professional education competencies (knowledges, skills, and attitudes) needed by beginning professional educators. These competencies serve as the foundation upon which each professional education course contributes to, and reinforces instruction within, the various professional course offerings comprising the professional education portion of the total curriculum. Course and content articulation is the ultimate goal in the design of a program of professional preparation. Such is in contrast to a series of unarticulated professional education courses.

The major difference between the CBTE and PBTE approaches is the performance in an actual classroom situation where students demonstrate mastery of the competency in question. As Cox states, in a competency-based teacher education program, the preservice professional program is designed to equip students with the professional competencies prior to student teaching (3.18). Performance-based teacher education connotes a type of professional preparation program whereby students develop a teaching skill and demonstrate the skill in an actual classroom situation. Consequently, the performance of the teaching competency in an actual classroom setting is the key factor in PBTE.

Competency-based teacher education programs are centered around a set of professional competencies which, if mastered, will enable one who demonstrates those competencies to be an effective teacher. In most cases the competencies are arranged in a sequential order to form an articulated professional program for the preparation of agricultural educators. Those professional competencies serving as a foundation for the preservice program are generally integrated into a "program package" or "professional competency core." The specific objectives for each professional education course are based upon specific professional competencies in order to provide continuity of purpose. From a curriculum standpoint, this approach helps eliminate unnecessary duplication and reduces overlap of course content.

Competency-based teacher education deals with specific teaching competencies in simulated situations where students are exposed to the various "teaching" competencies during their professional education coursework. Coursework and resultant exposure to the professional competencies are in preparation for application of these competencies in a real classroom situation at a later point in time. The key to CBTE is exposure to, and simulated practice of, professional education competencies. The performance of each competency may, or may not, be performed and evaluated in an actual classroom situation.

This is not to say that there is no application of those teaching competencies taught. Application occurs in micro-teaching demonstrations with peers. However, the actual performance, and resultant assessment, of the competencies taught in the preservice program occurs during the student teaching phase of the program. Cangelosi states that CBTE competencies typically are norm-referenced standards which preclude evaluation on a criteria basis (3.19). The integration of theory and practice of teaching in CBTE is not as great as for PBTE programs; but it is more than for course-based approaches.

The essential characteristics of the competency-based program are (3.20):

1. Competencies to be taught students are based upon what teachers must know and be able to do. The identified competencies become the course objectives of the articulated professional education phase of the preparation program.

2. Simulated experiences in pseudo-teaching situations are provided for each competency.

3. Students are exposed to, and practice, the competencies utilizing role playing, videotape recordings, micro-teaching demonstrations, and other methods in preparation for actual application.

4. Application and assessment of the competencies in a school situation occur during the time of student teaching.

Performance-based Approach

As mentioned earlier, the terms "competency-based teacher education" and "performance-based teacher education" are often used synonymously by educational leaders. This interchange of terms is due in part to the fact both program approaches stress mastery of professional education competencies. The word "competency" emphasizes that learning in both approaches is structured around identified and verified professional competencies needed by teachers.

Performance-based teacher education is an approach to teacher preparation in which each individual teacher is required to demonstrate essential professional competencies (teaching tasks) in a real classroom situation. Actual performance of the tasks ensures that the teacher has not only the necessary knowledge but also the ability to perform those competencies which are essential to successful teaching (3.21). Consequently, the key to PBTE is the teacher's demonstration of professional competence by performing the stated competency in a real classroom situation. The integration of teaching theory and practice is at the maximum in PBTE programs, since students must demonstrate their ability to perform as teachers in an actual classroom setting.

The essential characteristics of the performance-based program are (3.22):

1. Competencies to be demonstrated by students are identified, based upon what teachers must know and be able to do. Competencies are stated in behavioral terms which can be assessed objectively. They are shared with students at the beginning of the program. (Note: This characteristic would also apply to CBTE.)

2. Criteria to be used in assessing each specific competency are stated, including the conditions under which assessment will occur and the expected level of mastery. Criteria are also shared with the students at the beginning of the program.

3. Development and evaluation of the specific competencies by the students are focused upon in the instructional program.

4. The student's performance in the teaching role is used as the primary source of evidence in assessing the student's competency. Objective evidence of the student's knowledge related to planning, analyzing, interpreting, and evaluating true-to-life situations or behaviors is also considered.

5. The student's rate of progress through the program is determined by demonstrated competency rather than by time or course completion.

The National Center for Research in Vocational Education has developed a set of 100 modules designed for the implementation of performance-based vocational teacher education. Characteristics of this instructional system are (3.23):

1. Learning materials are developed and organized in the form of individualized packages, or "modules."

2. The program is designed to maximize individualized instruction.

3. The system is intended to provide students with immediate feedback after each learning experience.

4. Explicit criterion-referenced evaluation instruments are used to assess students' performances and progress.

5. Students are required to demonstrate in an actual school situation that they have achieved (mastered) the expected competency.

6. Most of the modules (learning packages) are self-contained.

Essential elements of the center's system required for implementation are (3.24):

1. Modules and related materials. The modules are units of learning, including a set of activities intended to facilitate the student's achievement of a teaching competency, specified in the form of a terminal behavior.

2. Resource person. A resource person is responsible for directing the student's progress through the program. This person could be a professor, cooperating teacher, master teacher, or supervisor.

3. Resource center. The resource center is an integral part of the PBTE program. It provides the student with access to most of the resources needed to complete the modules.

4. School setting. The program is field based rather than "classroom based." The final learning experience of all modules requires the student to demonstrate his/her competency in the actual teacher role; that is, to perform the competency in the actual school setting.

In summary, performance-based teacher education programs require performance in a real classroom situation; whereas competency-based teacher education programs tend to identify competencies and require performance but are not insistent that the final performance of each competency be in an actual classroom situation. That is, in a performance-based program, the learner is taught the competency, and that competency is applied in an actual classroom situation. In contrast, the learner in a competency-based teaching education program is exposed to (learns) the teaching competency, but actual application, or the performance of the competency in a true-to-life situation, may or may not occur at a later time.

A comparison of the three program approaches discussed in this chapter is presented in the following table:

TABLE 3.1--Characteristics of three program approaches

	Program Approaches		
Characteristics	Course-based	Competency-based	Performance-based
Objectives	Primarily theoretical	Primarily behavioral	Specifically behavioral
Content	Text or reference-based lecture/discussion	Competency-based	Competency-based
Nature	General cognitive understanding	Simulated/or real experience	Simulated performance and final performance in real situation
Evaluation	Objective-referenced	Norm-referenced	Criterion-referenced

(Continued)

TABLE 3.1 (Continued)

Characteristics	Program Approaches		
	Course-based	Competency-based	Performance-based
Feedback	Absent or group oriented	Immediate/ delayed	Generally immediate
Emphasis	Grades, credit, courses	Knowledge and/or demonstration of competencies	Demonstration of competencies
Completion	Time fixed cognitive testing	Time variable practice-based	Time variable performance-based

Program Duration

It is generally accepted that the undergraduate curriculum leading to a bachelor's degree in agriculture includes study in three distinct areas: general education, technical agriculture, and professional education. Regardless of the program approach utilized, the time involved to complete the undergraduate program is four or five years in duration.

Most programs have been designed in terms of a four-year duration. The first two years consist of coursework in basic agricultural sciences and general education. Students typically concentrate on their professional education courses in the junior and senior years, culminating in a student teaching experience in the senior year.

Over time, the major controversy among agricultural teacher educators has been the question of the most desirable division of students' time between technical agriculture, general education, and professional education coursework. As a means of increasing the amount of time available for students to master technical agriculture competence, some programs have been extended to five years. However, programs following this format have experienced difficulty in attracting students. Five-year programs have usually been unable to compete with four-year programs for students. Longer preservice preparation programs better equip teachers in the area of technical agriculture as well as in supervised field experience prior to entry into the teaching profession. However, due to the accelerating costs of higher education, loss of income, and increase in time required, programs in teacher education must out of economic necessity strive to prepare competent teachers within four years.

In recent years, many agricultural teacher education programs have initiated "new teacher" programs. These programs are designed to assist first-year teachers in planning instruction, managing the classroom, budgeting use of professional time, and developing total instructional programs of vocational agriculture. With the establishment of these types of "on-the-job" supervisory assistance programs, there is a less critical need for five-year preservice preparation programs. The new teacher programs in essence become the "fifth-year" programs. Many teacher educators feel that the supervision and follow-up of beginning teachers on the job is probably the most important aspect of their inservice education endeavors.

Program Design

Design denotes an outline showing the main features of something to be executed. Some of the major elements of program design which are essential for effective teacher education programs are addressed briefly in this chapter. However, no attempt is made to present a "model" program design for agricultural teacher education.

Instructional Programs

Irrespective of program approach employed, instructional programs in teacher education in agriculture generally include those in undergraduate education, graduate education, and inservice education. The ultimate intent of teacher education programs is to provide graduates with the necessary knowledges, skills, and attitudes essential so that they can contribute to the improvement of society as active and informed citizens, as educational leaders, and as teachers of youth and adults. The development of personnel who can also perpetuate the profession from a leadership standpoint is inherent in this intent.

Undergraduate Programs. There is a lack of unanimity regarding program content, scope, and procedures (methodology) in undergraduate agricultural teacher education programs in the United States. However, it is possible to identify some of the basic tenets which undergird the type of preservice programs essential to preparing individuals who are competent to plan, implement, and evaluate contemporary agricultural education programs.

Some of the more salient tenets are the following:

1. A balance of general, technical, and professional education is provided by the undergraduate degree program, regardless of the length of the training cycle.

2. Prospective agricultural educators are educated as professional in every sense. The development of, and subscription to, a sound philosophy of vocational education in agriculture is included.

3. The undergraduate program is based upon technical agriculture and professional education competencies essential for beginning agriculture teachers. The professional education competencies become the educational objectives for the professional education coursework.

4. Multi-track curricula which provide students with flexibility in specializing in various technical agriculture subject matter areas are developed by preservice education programs.

5. Undergraduate programs are designed, and adequately staffed, to prepare teachers of vocational agriculture. Undergraduates trained as teachers of vocational agriculture also become employed in many facets of professional agriculture.

6. The professional education aspect of the teacher education program consists of a series of articulated professional education courses designed not only to equip students with essential competencies but also to allow for the optimum philosophical and educational growth of the students.

The undergraduate instructional program is composed of coursework in the broad areas of general education, technical agriculture, and professional education. O'Kelley points out there is no great divergence among this nation's agricultural teacher education institutions concerning the inclusion of these broad areas in the undergraduate instructional program (3.25). He also mentions that most teacher educators do agree on the need for a strong technical agriculture core in the preservice program. Furthermore, most seem to agree on the need for general education study to the extent permissible within the time available. The major controversy in the profession has been to reach an agreement upon the most desirable division of a student's time between these three areas.

No attempt is made to prescribe the ideal balance between the preceding coursework categories. This issue was addressed by the agricultural education profession in 1976. As specified in the Standards for Quality Programs in Agricultural/Agribusiness Education:

> The agricultural education curriculum at the undergraduate level provides for adequate preparation of teachers of vocational agriculture/agribusiness, with course work distributed as follows:

General Education.	20-30%
Professional Education	20-30%
Technical Agriculture/Agribusiness	30-40%
Electives.	10%

A recent AATEA ad hoc committee evaluating coursework distribution revealed that 54.7 percent of the head teacher educators in agriculture felt that 29 percent of a student's program should be devoted to professional education coursework. There remains a great deal of diversity in opinion regarding the other three areas.

The exact percentage in each category is not the prime concern. The key finding is that there is a balance of coursework incorporated into the undergraduate instructional program.

Graduate Programs. Aspects of graduate education have traditionally been an integral part of teacher education in agriculture programs. Today, graduate education is an implied function of most agricultural teacher education programs. This facet of the overall instructional program has been designed to provide participants with the opportunity to increase both the breadth and depth of their professional and technical competence. In recent years there has been much emphasis placed upon creating and conducting graduate programs, especially at the Ph.D. level. The goal of these programs is to prepare individuals for leadership roles in agricultural education and for the broad overall area of vocational education at the state, national, and international levels in areas such as teacher education, supervision, administration, research, and curriculum development.

Structured formal graduate programs are designed for both the master's and doctorate degrees. Graduate doctoral-level programs are offered primarily in institutions where resources are available and employment possibilities exist in the state and region as well as in the nation.

Diversity in graduate programs exists between institutions of higher education offering advanced degrees. No attempt is made in this chapter to describe the various graduate programs offered in agricultural teacher education, since this topic is addressed in Chapter 12.

Effective graduate programs are characterized by the following (3.26):

1. Graduate degree programs provide for maximum flexibility in program design to meet the divergent needs, interests, and career objectives of students.

2. Graduate programs provide opportunity for an interface of practical and theoretical ideas, as well as communication beneficial to the overall growth and development of graduate students.

3. Training in applied research techniques is an integral part of the students' graduate programs.

4. Personnel leadership development is an implied part of each student's graduate instructional program.

5. Graduate programs emphasize the planning process. By capitalizing on existing abilities and incorporating research into education and technology, agricultural educators must make considerable effort to bring about improvements in local programs.

The hallmark of graduate programs in teacher education in agriculture over the years has been flexibility in program design, development of a strong research base, and emphasis upon personnel leadership development.

Inservice Programs. Inservice activities are designed and conducted to update and upgrade both the technical and professional competencies of agricultural educators in the field and, at the same time, provide opportunities for professional development. Credit and non-credit offerings in both technical and professional education, on and off campus, are offered by teacher education programs. In many institutions, it is possible to utilize credits generated through inservice endeavors to be applied toward an advanced degree.

A wide range of inservice activities are employed to aid practicing agricultural educators in improving their local education programs. Credit and non-credit short courses are offered on and off campus across the country. These courses are held during the summer, spring, and/or winter break and occur as workshops, special short-cycle summer school sessions, special problems, independent study, internships, and seminars. Beginning teacher courses, personal supervision, and follow-up are often the initial graduate experience. Teacher conferences are also used by agricultural teacher educators in offering inservice programs. The development and evaluation of curriculum and instructional resource materials provides additional opportunities for agricultural educators to improve their teaching effectiveness.

It is easy to recognize the importance of inservice education to aid in planning and conducting relevant and up-to-date instructional programs. Major responsibility in this area rests with teacher educators in agriculture. However, the need to plan inservice education programs cooperatively with teachers, teacher educators, and state staff members on a long-range basis is imperative.

In summary, inservice education is not synonymous with graduate education. Philosophically, they are not the same. Inservice education is directed primarily at the practitioner and designed to aid in improving education programs at the local level. In contrast, graduate education programs are more academic in design and intent. However, both inservice and graduate programs can be used to improve the overall performance of teachers in the profession.

Early Experience Programs

The background and experience of prospective agricultural educators has changed drastically in the last decade. A growing percentage of undergraduate students majoring in agricultural education have non-farm backgrounds. In addition, many students have not had the opportunity to learn about the organization and operation of vocational agriculture programs firsthand. Likewise, they have not gained personal experiences in participating in local youth leadership activities. Further, their practical experiences regarding occupations in agriculture are extremely limited. This trend will not likely be reversed in the immediate future.

Today, teacher educators cannot leave to chance the attainment of practical occupational and professional leadership experience development by prospective agricultural educators. Students will not gain essential and diversified experiences on the job. New instructional systems must be designed and implemented to assist future teachers in obtaining practical experiences. Consequently, early experience activities appropriate to the individual student's practical experience are integrated into his/her undergraduate preparation program. Bender supports the position that many students in colleges of agriculture and natural resources have an insufficient background of experience in agriculture (3.27). He further states that such experiences should be provided through occupational internships directed by the college.

Occupational Internships. Internship programs for the purpose of gaining occupational experiences have come into vogue in the last decade. They are designed and conducted primarily at the undergraduate level to provide an opportunity for students to gain firsthand experiences in one or more areas of technical agriculture. Internship programs are designed at the inservice level to strengthen the technical "know how" of practicing agricultural educators and to expand their industry experience. Occupational internships, at either the preservice or inservice level, are not to be confused with five-year programs, professional internships, or externships which are used in various teacher education programs.

Not only do such programs provide an opportunity for students with limited occupational background to gain practical experiences in production agriculture and management, but they also give students early exposure and involvement in vocational agriculture programs at the local level. For early exposure/experience programs to be effective, educational plans and placement agreements have to be developed. Internship programs include a formal placement agreement between the student intern, personnel from the institution, and the employer-cooperator. Likewise, educational plans are being developed which specify the tasks (knowledges and skills) to be achieved by the student intern and which are tailored to the individual needs of each student.

Leadership Development Internships. Another kind of early experience program being used in agricultural teacher education programs provides practical leadership experience. This type of internship provides students with leadership skills. Agricultural education majors are placed in a local school program utilizing an individualized and open-entry/open-exit concept. The internship is designed to provide students, especially those who lack leadership training experiences, with the opportunity to participate in leadership development activities in a local chapter of the Future Farmers of America. This activity also provides early exposure and involvement with a local vocational agriculture program prior to student teaching. As with occupational internships, the leadership development internships are integrated into the undergraduate program for maximum benefit to students.

Professional agricultural education courses are usually concentrated in the junior and senior years. Because of the structure of teacher education in agriculture programs at the preservice level, all avenues must be explored to give prospective agricultural educators early exposure and involvement in working with students, as well as providing the opportunity to gain practical experience in agriculture.

The Block Concept

Teacher education in agriculture programs must be planned to provide an articulated instructional endeavor whereby students can progress from one level of instruction to the next and, at the same time, acquire new and broadened competencies essential for their successful entry into the teaching profession. The instructional and experience content of the preservice program should be designed so that each objective, competency, and professional education course builds upon another.

To facilitate the fusion of content and experience throughout the preservice program, program articulation is essential. Articulation is the process used to provide the quality, quantity, and variety of educational activities needed to move students from one educational level to another without undue overlap. It is through the mechanism of articulation that teacher educators have greater assurance that students are being prepared with those competencies essential for entry into the teaching profession. Similarly, that entry is made with some degree of success.

The first two years of the undergraduate program in agricultural teacher education usually consist of coursework in basic agricultural science and general education, which includes biological and physical sciences, mathematics, social sciences, communications, and humanities. It is generally agreed that general education courses should be undertaken early to provide a background for the applied courses.

In the junior year, students usually enroll in general professional education as well as agricultural education courses. Advanced-level technical agriculture coursework is also a part of the junior experience. During the senior year, students complete the required professional agricultural education and additional technical agriculture coursework. Student teaching is usually completed during the senior year.

Professional education courses are generally concentrated in the last two years of the preservice program. Many different procedures are utilized to expose and provide students with the pedagogical skills essential for students to have a successful field experience (student teaching). In the final analysis, field experience is the focal point of the preservice program.

Some institutions concentrate upon providing students with only "survival skills" prior to student teaching. The belief of teacher educators employing this avenue of teacher education is that prospective teachers learn more and better how to teach when they are in an actual classroom setting. Proponents of this procedure believe future teachers will master those essential teaching competencies on the job. For example, the best way to learn how to prepare a lesson plan, how to use oral questioning, or how to conduct an effective field trip would be during the student teaching period under the supervision of a cooperating teacher. Those who expound this methodology believe "trial and error" is the best approach to developing teaching competence. Survival skills taught prior to student teaching serve the purpose of equipping students with sufficient skills so they can master essential teaching competencies on the job. Following field experience, students return to campus to study methods of effective teaching and to analyze why events occurred the way they did during student teaching.

Other institutions preparing prospective agricultural educators provide all the professional education coursework in specific semesters or quarters prior to student teaching. An exception is the use of post-student teaching seminars to expose students to additional professional competencies, such as department administration and program planning. These seminars also provide students with the opportunity to share student teaching experiences and, in some instances, to analyze specific problems.

Proponents who use this procedure to provide students with competencies essential for successful entry into the teaching profession believe that theory should precede practice. They advocate that students learn better during supervised field experience if students are knowledgeable about methods of teaching and applied educational psychology prior to student teaching. It is much easier to extend the student teaching time to encompass one full quarter or one full semester with this avenue.

Program articulation is an essential element for a meaningful experience in the education of teachers. To promote maximum articulation, many agricultural teacher education programs employ the "block concept." The "block" program, sometimes referred to as the professional semester or quarter, culminates the preservice preparation program. The block concept combines student teaching with various professional education courses, such as "Methods of Teaching Vocational Agriculture." This concept attempts to integrate the theory of teaching with actual teaching experience during student teaching (usually in the senior year) by concentrating the two dimensions (theory and practice) into a coordinated, structured block of time.

The block concept is widely used in teacher education in agriculture programs, and there are many variations of this concept in use today. However, the essential components of the block concept may be identified as follows:

1. Early experience in the training center prior to the block semester or quarter.

2. Professional agricultural education courses of an intensive nature taught the first part of the professional semester or quarter.

3. Full-time student teaching immediately following the professional education coursework.

4. Post-student teaching seminars.

The problems inherent in the implementation of the block program include:

1. Scheduling and administering the program when students are away from campus for as long as 10 weeks during the academic year.

2. Understanding the added cost to students created by living in a community, usually away from campus, on a full-time basis.

3. Compressing a large proportion of the preservice professional competencies into one quarter or semester.

4. Providing adequate faculty supervision of student teachers who are located throughout the state.

Even with the inherent problems, the block concept still seems the best way for "concentration of purpose." Preservice programs of teacher education, which include an early contact with the teaching situation, a period of intensive study (professional education coursework) on campus, a period of full-time student teaching, and, finally, a return to campus for post-student teaching seminars, closely parallel the logical steps in problem solving; that is, the problem is identified; relevant data are examined and plans are made; and the plans are applied to the solution of the problem, and their effectiveness is evaluated (3.28). The experience has unity. There is fusion of theory and practice. To facilitate this fusion, students are completely immersed in full-time resident field experience in the selected training center.

Field Experience Participation

Field experience is the heart of the preservice preparation program, since the integration of theory and practice is an essential aspect of an effective teacher education program. O'Kelley succinctly points out that the most important aspect of the supervised teaching program (student teaching), if not its only justification for existence, lies in the kind and the scope of teaching experiences obtained by the student teacher (3.29). He further mentions that the mere observation of teaching activities is not sufficient. The opportunity to gain actual experience in typical teaching situations on a continuing basis is of paramount importance.

Few teacher educators in agriculture would disagree that professional field experience is the most satisfying and productive facet in the preparation of agricultural educators. However, ample planning, effective guidance, and wholehearted counsel of student teachers during this critical period are essential to the success of the agricultural teacher education program. Cooperating teachers are the key individuals since they set the future patterns, habits, attitudes, and abilities of prospective agricultural educators. Without exception, the quality of the cooperating teachers determines the quality of the student teaching program. The impressions made on the student teachers by the cooperating teachers may well be more permanent than those made by teacher educators.

For the field experience (student teaching) phase of the preservice preparation of agricultural educators to be effective for the student, there must be a close coordination between the activities experienced in the training center and the professional education competencies taught in teacher education programs, regardless of the program approach utilized. In order to reinforce the basic principles and professional competencies taught prior to student teaching, the student teacher must be afforded the opportunity to apply these concepts to real-life situations. What is taught on campus in the preservice programs must be experienced during student teaching. In essence, there should be integration of professional education competencies into the field experience program if the fusion of theory and practice is to be accomplished.

One unique characteristic of student teaching in agricultural education is that, throughout the country, teacher educators themselves are the university supervisors. Not only do they teach the professional education courses, but also they participate actively in student teacher supervision and evaluation. This feature provides an opportunity for the teacher educators to see their instruction being applied in actual classroom situations.

Similarly, the preservice preparation program should be examined in order to assure that prospective agricultural educators are receiving the kinds of technical and practical agriculture instruction required for contemporary programs in agriculture.

Technical and Practical Instruction

The technical agriculture facet of the preservice program provides students with the opportunity to acquire the technical competence necessary to conduct effective instructional programs in agriculture for youth and adults.

With the passage of the Vocational Education Act of 1963, instructional programs in vocational agriculture began to undergo a transition from strictly production agriculture to encompass also agribusiness and natural resource occupations. Agricultural educators from the local to national levels have the responsibility for planning instructional programs to meet the needs of persons who will be engaged in production agriculture, as well as those engaged in agriculturally related occupations. As a result of the 1963 act, the era of program specialization was born. Seven instructional taxonomies emerged, and they are accepted today as major areas of program thrust. Specialized programs of instruction are designed to provide greater indepth knowledges and skills in a cluster of agricultural occupations inherent within one of the seven program taxonomies.

At least three definite program features have evolved in recent years to facilitate the acquisition of technical competence by future teachers. These features are (1) emphasis upon agricultural science principles and competencies, (2) dual-major options, and (3) practicums.

Emphasis upon Agricultural Principles. Future curricula in technical agriculture must provide greater emphasis on basic agricultural science principles and their application to the real world. It is imperative that courses in technical agriculture for future agricultural educators stress basic scientific concepts, principles, and relationships germane to an indepth understanding of all major fields of agricultural subject matter (3.30).

The need to generate technical agriculture courses which will provide opportunities for students to apply agricultural theory, knowledges, and skills has far-reaching implications for agricultural teacher educators. In most instances, course offerings in the various subject matter departments in the college of agriculture are utilized to provide the technical agriculture instruction needed by future teachers. Few, if any, agricultural teacher educators can dictate the course objectives, competencies, and content offered in these technical agriculture service courses. Thus, it is essential that a cooperative relationship exist between teacher educators and the departments providing technical service courses to future agricultural educators.

Attempts have been made in a few institutions to identify a set of core competencies in basic agriculture, which would be appropriate for all students in agriculture regardless of majors (3.31). The underlying purpose is to provide graduates from the college of agriculture with those competencies needed to discharge their duties and responsibilities satisfactorily. It is assumed that once these core competencies are identified and validated, they will be used by instructors of service courses in technical agriculture to develop course outlines, which tends to assure that the important knowledges and skills are taught. Some degree of assurance of the relevancy of course content is thus provided. Ultimately, the core competencies can also assist in providing articulation between technical agriculture courses so as to avoid duplication of instructional content.

Dual-Major Options. In the last decade, the use of dual majors became an essential dimension of many undergraduate programs preparing future agricultural educators. Dual-major options provide students with the opportunity to gain greater specialization in selected areas of technical agriculture. In essence, prospective agricultural educators can enroll in dual-major programs--agricultural education plus a technical agriculture field.

Adequate and detailed planning of students' courses of study is required if students are to complete requirements for two majors in four years. This program feature has served as an effective recruitment tool, since many students are attracted by this type of program flexibility. Dual-major programs tend to satisfy student interests to a greater degree than a one-track curriculum, and they also provide an effective avenue to secure more specialized training in technical agriculture. The concept of dual majors also helps meet the demand for specialization in one or more of the instructional taxonomies.

Practicums. Curricula in colleges of agriculture must provide students majoring in agricultural education with practical applications of various skills, as well as with cognitive knowledge. Bender states that a long-founded principle widely held by agricultural educators is that teachers should be able to operate from the perspective of the real world (3.32). He further points out that principles should be taught and understood but that they can best be understood if students are afforded the opportunity to engage in the application of the principles.

The use of practicum-type courses provides an excellent avenue to integrate theory and practice of agricultural science principles and concepts. Teacher education programs in agriculture have employed the practicum concept for years in teaching agricultural mechanics skills to prospective agricultural educators. Many programs employ agricultural mechanics specialists who plan and teach courses to serve the needs of students in this important area of technical agriculture.

Likewise, the need in the future for practicums in other areas of technical agriculture will be even greater than in the past. As more students with non-agricultural backgrounds enter agricultural teacher education programs, practicum-type courses must be augmented. It is estimated that from one-half to two-thirds of the students who enroll in agriculture colleges do not have a farm or ranch background or significant agricultural production experience. The lack of practical experience on the part of contemporary students enrolled in high education dictates the need to incorporate basic agricultural skills into the overall program.

In summary, programs of teacher education in agriculture must assume new responsibilities not inherent in preservice programs of a decade ago. Future programs must include specialization and practical application of agricultural science principles in addition to providing knowledge of these principles. More practicum courses must be incorporated into the courses of study for future agricultural educators if they are to be prepared to implement "learning-by-doing" instructional programs in a constantly changing agricultural environment. The teacher of vocational agriculture must have practical production agriculture experience to relate effectively to farmers and ranchers.

Extension Education

The future demand for versatile agricultural educators has already been expressed in the light of changing instructional thrusts for contemporary agricultural industry. An agricultural educator has been broadly defined as one who has extensive preparation in agricultural disciplines and knowledge and understanding of the teaching-learning process (3.33). If one accepts this broad definition of an agricultural educator, then one must also accept that preparation of cooperative extension workers lies within the realm of teacher education in agriculture programs. Extension personnel are educators functioning primarily in an informal educational arena. There are many similar professional competencies common to both agriculture teachers and agricultural extension workers. To be effective, future extension workers will need much preparation in both technical agriculture and professional education.

Bender stresses that it is very important for students interested in preparing for careers as professional extension workers to be prepared both technically and professionally in colleges of agriculture (3.34). Programs of agricultural education appear to be the best source of this preparation, since there is an overlap of the knowledges, skills, and attitudes essential for both types of agricultural educators. However, there is not unanimous agreement between agricultural teacher educators as to where preparation programs in extension education should be located. This issue has been debated by Bender and Crawford (3.35).

If teacher education in agriculture programs are to be as dynamic as the agricultural sector they serve, the preparation programs must reflect the reality of this sector. By definition, curricula designed to prepare contemporary agricultural educators must be based upon the competencies (skills, knowledges, and attitudes) these educators need to plan, implement, and evaluate high quality educational offerings effectively. A multi-track preparation program encompassing both core competencies and specialized professional competencies unique to agricultural teaching and agricultural extension warrants consideration. This multi-track program allows for (1) greater efficiency in the use of learning resources and (2) considerable cross-fertilization between vocational teaching and extension.

If a broader definition of agricultural educators is accepted, university programs of agricultural education must assume increased responsibility. In the final analysis, future efforts must encompass professional preparation and competency development for all persons involved in educational processes associated with the broader sector of agriculture.

Other Concerns Affecting Programs

The future of teacher education in agriculture programs holds potential for many new and demanding challenges. However, in all probability, new directions and responsibilities will emerge in the following areas:

1. Expanded supervision and administrative services to local education agencies (LEA's). There will be more contractual services offered by teacher education to LEA's in the future.

2. More inservice education to meet the needs of a diversified group of agricultural educators. Both technical and pedagogical competency development will be involved.

3. Renewed emphasis upon program development, long-range program planning, and comprehensive program evaluation for local education agencies. Programs of teacher education will contract to conduct program evaluation for state departments of education.

4. Increased involvement with state vocational affiliates to affect both state and federal legislation for vocational education. A further decline of vocational funding for teacher education, coupled with the loss of categorical funding for agricultural education, will force teacher educators to become more involved in the legislative process.

5. Expanded research efforts which are identified and funded by state departments of education. More emphasis will be placed upon contractual services.

6. Greater involvement in international education programs. This area will receive more emphasis from college administrators as college enrollments decline.

7. Increased involvement in resolving certification issues such as temporary, course-oriented, competency-based, and/or performance-based teacher certification.

8. Greater emphasis in developing reading, communication, and computational skills on the part of agricultural educators in the future.

The preceding list is not complete. However, it should serve to point out new and expanded responsibilities teacher educators in agriculture can expect to inherit in the future. This is as it should be. Teacher educators in agriculture have always provided the leadership essential for keeping education in agriculture dynamic and viable. Dynamic leadership, which is continually seeking better avenues for program effectiveness, must continue.

Summary

This chapter has outlined the development of teacher education in agriculture programs in the light of functions, principles, and guidelines. Particular attention has been directed to updating the development of programs of teacher education in agriculture since 1967. Teacher education objectives and responsibilities have been discussed in an attempt to focus upon the increasing complexity of demands placed on the profession.

The processes and procedures used in the identification, legitimization, and implementation of program standards affecting teacher education programs in agriculture have been identified.

A discussion of program design has been presented in an attempt to strike a balance between what has been, what is, and what should be considered essential. Various program approaches utilized in teacher preparation programs today have been highlighted. Several essential ingredients or effective teacher education programs have been discussed. These include utilizing early experience programs; using block concept to integrate theory and practice of teaching; providing participating field experience, along with technical and practical instruction; and advocating practical hands-on instruction.

The chapter culminates with a projection of new directions and responsibilities for teacher education in agriculture programs.

CHAPTER 4

Instructional Objectives for Preparing Teachers

Max L. Amberson
MONTANA STATE UNIVERSITY

Douglas Bishop
MONTANA STATE UNIVERSITY

It has been said that history repeats itself and that history is one of our greatest teachers. Assuming that both tenets are true, teacher educators in agriculture, prospective vocational agriculture teachers, and others should be able to learn much from the evolution and application of the educational processes involved in teaching vocational agriculture. But, this approach could lead to some misinterpretations of the role of vocational agriculture teachers, particularly if there have been any contradictions throughout the historical development of vocational agriculture. At any rate, history, regardless of its clarity, defies simple interpretation.

A more defensible approach to discussing the role of teachers of vocational agriculture would be to identify the current objectives of teacher education in agriculture first and then compare the final product with the established objectives. One dictionary definition of an "objective" is that it is ". . . a known or perceived object as distinguished from something existing only in the mind (4.1)." This would imply that the measure of our success is predicated upon a planned program and the result of our efforts. With such a background, the future of agricultural education is more predictable.

Teacher educators are often asked to assess the present state of teaching vocational agriculture and to synthesize the assorted views of what the future holds. If we are to accomplish such a formidable task, we must make two basic assumptions. First, teacher education in agriculture is a means to an end and not merely an end in itself. Second, the objectives of teacher education in vocational agriculture education are deeply rooted in the time-tested philosophy of effective teaching and learning.

It is perhaps easiest to assess the role of teacher education in agriculture by first looking at the role of the teachers we prepare. It should be recognized, however, that the role of the vocational agriculture teacher has changed and will continue to change. While the agriculture program was once almost a separate unit within an educational system, it has now become a highly articulated facet of a very complex educational endeavor. Today's successful vocational agriculture program is no longer isolated in its own domain, for the program has become an integral part of the total educational environment in both rural and urban communities. Within this context, the secondary agriculture teacher will continue to play a prominent role in the future.

It has been established that of all factors operating within the classroom, the teacher is the most influential factor in terms of the student's learning. Thus, today's teachers must recognize their vocational as well as their non-vocational role in preparing workers for the future.

Teachers of Vocational Agriculture

What then must teacher education prepare teachers to do? Vocational agriculture teachers must increase every student's sense of personal worth as well as prepare each one for entry-level occupations. By searching for that knowledge or skill that every student can be proud of, teachers can build up their students' self-confidence. Students have little difficulty seeing through a posture of moral or intellectual superiority.

As the role of the vocational agriculture teacher continues to change and expand, and as continuing change occurs in agricultural technology, more subject matter specialization will be required. Such advancements have resulted in the need for a cadre of teachers whose preparation program may include production agriculture, horticulture, or power mechanics. The range of competencies required by agriculture/agribusiness employees increases in proportion to the overall increase in agricultural technology. Thus, today's teachers are faced with the reoccurring task of identifying what it is they must teach.

The increase in technology and the subsequent increase in the specialized competencies required by agriculture employees has resulted in more multi-teacher departments. This is especially true in the larger urbanized areas where the division between production agriculture and agribusiness is more evident. In this setting, teachers are more apt to have organized their programs into semester courses, modular programs, and concentrated blocks of instruction to accommodate the individual needs of their students and the local industries. Since many students do not have backgrounds in production agriculture, it has been necessary to implement core programs to assure mastery of basic common competencies essential for job entry into both agricultural production and agribusiness.

Although specialization may seem to hold the answer to many problems facing agricultural education, large numbers of one-teacher departments still exist. Many of these are located in sparsely populated areas where there is not a large number of agribusinesses. In these cases, more by necessity than by current trend, teachers remain generalists. To provide for specialized programs, teachers must call upon outside resource persons to give specialized information and assistance to themselves and their students.

The student population is also changing. A smaller percentage of students have knowledge and experience in agriculture/agribusiness, even though they may have a broader general knowledge of many other topics.

Aside from the changes in technology and the differences in students' backgrounds, teachers must face a public which often does not appreciate or understand the role of agriculture in this country. Fewer and fewer individuals in many of our communities are aware of the complex nature of agriculture and its importance as the basis for a healthy economy. Many individuals are questioning many of the time-tested tenents upon which vocational agriculture teachers based their programs in the past. Having learned to expect an abundance of agricultural products, they are often guilty of assuming such abundance has always been and will continue to be. Therefore, they fail to see the need to educate the populace for occupations which may be of high interest. They fail to realize that, in part, production agriculture and agribusinesses have become dynamic, efficient segments of our economy because of the education traditionally provided when preparing their work force.

The teacher education program offers an opportunity for men and women to stay abreast of the rapid changes in agriculture and to work with agricultural producers/agribusinesspersons and their families. Today's vocational agriculture teacher typically has a minimum of a bachelor's degree. Along with the more traditional technical courses in agriculture and mechanics, courses in natural sciences, mathematics, humanities, social sciences, communications, education, and fine arts are included. Such academic preparation supported by work experience in agriculture results in a unique kind of teacher.

Current teacher education programs are preparing teachers for a critical period in the evolution of U.S. agriculture. Our society is changing faster than ever before. New agricultural technology, greater application of scientific principles in agriculture, bigger, more complicated machines to do our work for us, and more community interaction tend to increase the complexity of agriculture. Concurrently, there are social and attitudinal changes that hold implications for the preparation of teachers of vocational agriculture. Teacher education must change to meet the new demands of those it serves if it is to continue to be a viable part of the public service facet of our society.

Characteristics of Effective Agriculture Teachers

What makes a vocational agriculture teacher different from other teachers, if in fact they are different? All good teachers possess contagious enthusiasm, are innovative, care about students, and are concerned with the interests and welfare of the community. Any differences that exist lie in the various categories of tasks teachers of vocational agriculture are expected to perform in their local communities and not in the array of personal and professional characteristics they possess.

A teacher of agriculture often enjoys more freedom than teachers in academic subjects in establishing the goals and curriculum for the program. The diverse nature of agriculture from community to community requires that each teacher tailor the program to a specific situation. The freedom to determine the nature of the program is often further brought about because vocational agriculture may be physically separated from the rest of the school. Although such separation often gives teachers more freedom, it puts an additional burden of responsibility on them to communicate effectively with the rest of the school faculty and administration.

Vocational agriculture teachers throughout the United States are recognized for their practical approach to education, their keen interest in meeting the needs of their students--both youth and adults, and their concern for their communities as well as their schools.

Teachers of agriculture, in many cases, are liaisons between their schools and the agriculture/agribusiness communities. The teachers' daily association with local persons in agribusinesses, cooperative extension, banks, and county government, as well as with individuals in farming or ranching and the various farm organizations, places them in a unique position to lend support to the educational system within their respective communities.

The knowledge of, and close association with, their communities enables agriculture teachers to identify the occupational needs of their communities and to interpret these needs to the students, parents, and employers. Teachers of agriculture find themselves using local resources as tools in the classroom to make learning interesting for the students and to locate jobs whereby the students can gain supervised occupational work experience.

Because the activities of vocational agriculture teachers frequently take them into their communities, their role as special ambassadors of their schools is significant. Frequent teacher contacts with community businesspersons and organizations provides for an exchange of information, while adding to the uniqueness of their role as secondary teachers. In addition, numerous contacts in their communities enable agriculture teachers to build a favorable image of their schools in the minds of the students and in the minds of the public. The teachers, to be effective, must establish rapport with their respective communities as well as with their students.

Vocational agriculture teachers are often hired on an extended contract--either 11 or 12 months, which makes them year-round employees of their schools and communities. The importance of the extended contract cannot be overemphasized. By being employed on either an 11- or a 12-month contract, they have additional time during the summer to plan for program improvements, supervise students' occupational experience programs, and act as advisors to the FFA. Due to the seasonal nature of agriculture/agribusiness, many of the related experiences necessary for the adequate education of the students occur during the summer months.

There are four major identifiable facets to the role of the vocational agriculture teacher, each of which will be described in some detail later. These major aspects are (1) providing classroom/laboratory instruction, including instruction in agricultural science, leadership, and mechanics, (2) coordinating supervised occupational experience (SOE) programs, (3) advising the FFA, and (4) instructing out-of-school youth and adults in production agriculture/agribusiness.

These four major areas have constituted the teacher job description since the beginning of vocational agriculture; the role of the teacher in accomplishing these tasks however continues to change.

Early in the development of the vocational agriculture program, the vocational agriculture instructor and the county extension agent were the local experts in matters dealing with technical agriculture. More recently, many public and private agencies have been providing technical information and assistance directly to farmers. This change, along with the increased educational level of today's farmer and rancher, has resulted in the local agriculture teacher becoming a facilitator, coordinator, and/or an interpreter of technical knowledge within the community. The agriculture teacher nationwide has become less of a generalist and more of a specialist.

This change in role and responsibility tends to increase the educational requirements for teachers of agriculture. Where the bachelor's degree was once sufficient, many teachers of agriculture are now involved in participating in intensive inservice programs and in obtaining their advanced degrees. The continual upgrading of the educational level of the farmer and rancher is having far-reaching effects on the role of vocational agriculture instructors.

Competency-based Teaching

Production agriculture/agribusiness is such an all-inclusive subject that vocational agriculture teachers often have difficulty in identifying the content of their programs. The classes may be organized into the traditional Ag I, II, III, and IV year-long classes or in units or semesters to present skills, knowledges, and attitudes needed for successful employment. In response to this situation, most state curriculum guides have evolved around a competency-based approach.

From the very beginning of vocational agriculture programs, effective teachers have been concerned with performance-based instruction. The effective vocational agriculture teacher has always helped young people establish goals and objectives for each facet of the agriculture program. In addition to student development of competencies, monetary return from the student occupational experience program has also been considered a valid measure of instruction competencies taught in the program.

The emphasis on competence-based education will cause vocational agriculture teachers to become more systematic in their instruction. Certain accomplishments may be evident in the classroom and throughout the program, reflecting that a teacher has developed the program utilizing the competency-based concept. Effective use of the competency approach is reflected in the program development process as follows (4.2):

1. Employment opportunities for the present and future are identified, based on systematically collected local, state, and national data.

2. A task listing of what agriculture/agribusiness workers do on the job is reflected in the curriculum.

3. The identified tasks are studied carefully to determine their relevance to the needs and the specific level of the individual student's program.

4. Performance objectives are then written to describe accurately the student performance that will be acceptable for the needed tasks to be executed.

5. Detailed lesson plans are developed to enable the teacher to present the course content systematically.

6. Once the teaching plans have been prepared, appropriate materials and/or media which can be adapted to the performance-based approach

to learning must be found. In some cases, material and/or media must be developed to fit the individual teacher's plans.

7. The lessons are taught, student performance is measured, and subsequent revisions are made to improve the teaching-learning environment.

8. A periodic update of the task analysis is conducted to determine any changes in the tasks performed by workers being trained.

Preparing teachers to use competency-based instruction in vocational agriculture has resulted in some changes in emphasis in the teacher-training programs. However, instead of emphasizing subject matter content, teacher preparation programs must continue to emphasize the development of local programs which evolve around the interests of the students and the needs of the communities.

1. Teachers must have a sound philosophy for teaching vocational agriculture, one which revolves around the individual student.

2. Teachers must become "directors of learning" rather than the traditional disseminators of knowledge and skill.

3. Teachers must become more familiar with evaluation procedures and techniques to assure that they are measuring and reporting student performance accurately.

4. Teachers must learn to develop both short- and long-range goals and must have actual experience in developing and using competency-based instructional materials.

Inschool Instruction

A major responsibility of teachers of agriculture is to plan, teach, and evaluate the learning outcomes of their students. As in the past, they are continuing to use primarily the classroom and laboratories available to them, which may include a mechanics area, a greenhouse, a school farm, and a livestock facility. Teachers of agriculture, unlike other teachers in the school, perceive their classrooms broadly, that is, the total community. Like other teachers, vocational agriculture teachers are taught how to make teaching plans, prepare tests, give assignments, and evaluate students' work--activities common to all teachers. All teachers are responsible for individual classes, but the vocational agriculture teacher is responsible for planning and conducting a wide variety of instructional activities related to agriculture and agribusiness.

Individuals outside agricultural education often do not understand the meaning of a complete program of activities. Rather than preparing students for entry into the world of work and for advanced education, they may equate inschool teaching with a given textbook on a specific topic for a period of time while the students are confined within the four walls of the school. However, the vocational agriculture teacher must combine many topics from numerous texts and other resources into a meaningful learning package that will prepare students for entry into the world of work, since many students terminate their formal education following high school.

Unlike many traditional courses, vocational agriculture is on a voluntary, or elective, basis. Some students take the course because they are inherently interested in agriculture, while others take it just because it is an elective course. This free-choice option has implications for classroom instruction in that instructional content needs to be interesting and useful, and it needs to promote thinking on the part of the students.

Effective instruction in the classroom and laboratory requires that teachers of agriculture assess the needs and expectations of the employers in the many types of agriculture/agribusiness occupations in their respective communities. Keeping in touch may require additional time above and beyond that spent by many teachers in other fields. A vocational agriculture teacher's typical day is not completed when the bell sounds at the end of the last period of the day.

Instruction in vocational agriculture is not confined to the school. Teaching may take place in a field of diseased corn, at the local livestock commission yard, or at the local agricultural machinery dealership. Agriculture teachers use their communities and their resources to create an effective environment for teaching. They recognize that meaningful instruction requires a realistic learning situation.

In many cases teachers cannot take students to resources in their communities. It is often necessary to transport community resources not normally found in the school to the school to be observed, studied, and used in order to provide effective learning experiences. A myriad of feed samples, a digestive tract from the local slaughter plant, the latest tractor, a local legislator are but a few examples of what a community can provide in the way of teaching materials and resource persons. Teacher educators should set the example by utilizing all the services available to them during the preservice preparation of agriculture teachers.

Coordinating an effective inschool program within a school system is no small task. Effective vocational agriculture teachers must utilize well-planned objectives as a guide to both short- and long-range instructional planning.

The task of the teachers becomes even more complicated when one considers that the subject matter taught in agriculture must relate to, and utilize, all other disciplines--for example, the use of math when balancing feed rations, grammar when writing business letters, and physics when designing farm implements. There is a growing need and demand for agriculture teachers to teach introductory agriculture material to the elementary grades. This area of responsibility needs further development as fewer and fewer persons are growing up with any association with production agriculture/agribusiness.

To help accomplish the task of establishing instructional priorities that meet the needs of their students and communities, agriculture teachers must develop and test specific instructional objectives.

Well-defined teaching objectives that are performance-oriented have been characteristic of vocational agriculture and have been used as a means to implement program goals for many years. The amount and nature of progress made in the instructional program can be measured by the extent to which stated objectives are accomplished. Gentry, writing in Agricultural Education in 1937, stated, "While ultimate goals are of assistance in determining direction they are not of much help in determining rate of progress or growth. It is necessary to set up intervening mile posts in line with our major aims. These mile or quarter mile posts we will call 'objectives' . . . (4.3)." Certainly, the concept of performance objectives is not new to vocational agriculture. Much research has been done in an effort to refine the concept of performance objectives in all areas of education. Because of the many different programs that are being developed today, teachers of agriculture should be prepared to establish goals and measurable objectives that can be used to direct the learning process in the classroom. Perhaps the best way to prepare secondary teachers in the use of instructional objectives is to organize teacher education programs on a similar basis. Professional courses in which all agriculture teachers enroll would provide a good example for them to follow when they become teachers.

The National Agricultural Occupations Competency Study (4.4), completed in 1978, should provide considerable assistance in developing vocational agriculture programs. The project, designed to identify the essential agricultural competencies needed for entry-level employment and advancement in the major agriculture/agribusiness occupations, will serve as a guide for future instruction in vocational agriculture.

Teachers of vocational agriculture must continue to conduct student-centered instruction. Future classroom teaching will probably involve numerous small groups and, in many cases, individuals. The increased diversity of occupations in the agricultural sector and the rapid change in entry-level job requirements for agriculture/agribusiness will require well-qualfied, highly specialized teachers of vocational agriculture.

Supervised Occupational Experience

The orientation of school personnel, students, parents, and employers to the role of supervised occupational experience (SOE) programs in vocational agriculture is essential. The question which has been posed by those outside the field is "Why are SOE programs an essential part of the vocational agriculture/agribusiness program?" Agricultural educators should strive to develop the level of awareness to a point where the public perceives supervised occupational education as an integral part of the student's program such that the question becomes "What is the best type of SOE program?"

Although vocational agriculture/agribusiness is an elective subject and participation is voluntary, for students in the program, FFA and SOE programs are not voluntary, for they are considered essential in preparing students adequately for employment.

SOE programs apply the learning-by-doing principle, which is a proven method of instruction used since the beginning of vocational education. Currently, most state plans for vocational education indicate that all students enrolled in vocational agriculture will have supervised occupational experience as a part of their instructional programs. The responsibility for implementing this policy rests squarely on the shoulders of the local teachers. In cases where SOE is used successfully, the teachers systematically direct students in planning experience programs which can lead to the development of the knowledges, skills, and attitudes needed for entry-level employment and advancement in agriculture/agribusiness jobs.

With the completion of competency studies at both the state and national levels, teachers have ready access to references that can be adapted to local situations for identifying job competencies needed by prospective employees. These references are particularly important for use by the teachers, students, and employers in tailoring the students' SOE educational plans to be followed while the students are working on the job.

In order for teachers to help plan appropriate SOE programs, they first must know the career goals of the individual students enrolled in vocational agriculture/agribusiness. The question to be asked is how can the teacher best help students prepare to master the appropriate knowledges, skills, and attitudes they need to enter and advance in their career choice in agriculture/agribusiness? Thus, a student's occupational interest is the key to designing that student's SOE program.

There are no standard answers in organizing SOE programs, since there are many different approaches, occupations, and student backgrounds. The experience programs for two students headed for a career in agricultural beef production may be quite different if one was reared on a commercial beef ranch and the other on a one-acre lot in suburbia with no previous ranch/farm experience. Thus, no one type of SOE program is "best," and the variations just described have led to the evolution of four general configurations of SOE programs that are being used successfully. These are ownership, placement, cooperative, and school laboratory.

Ownership

Most individuals refer to the ownership arrangement as a "project program," primarily for obtaining experience for self-employment in agricultural production. In this program students own and manage the resources that go into production enterprises. The program's basic values are that it gives students pride of ownership and it helps them appreciate the need for management experience. By taking part in ownership programs, students may grow into entrepreneurs. Although the ownership SOE program is generally associated with production agriculture, the idea is equally applicable in the other taxonomy areas in agribusiness. Students are currently using ownership SOE program concepts successfully in horticulture, mechanics, production agriculture, and forestry.

Placement

Placement in production agriculture/agribusiness is not a second choice to ownership. Rather, for many students it is a more appropriate means for employment in an agriculture/agribusiness occupation. Initially, a placement program can provide career exploration in the field in which a student feels he/she has an interest. Even for students with a career goal of self-employment, placement is often the most feasible way to gain experience, due to an initial lack of capital or opportunity to pursue ownership SOE programs.

Cooperative

Cooperative-type SOE programs, specifically named and funded in the Vocational Education Act of 1963, often resemble placement SOE programs except that they tend to be reserved for students who have broad general preparation and experience in agriculture and are seeking indepth experience in a somewhat narrower area. Cooperative programs are often called capstone experiences, since they "cap-off" the final experience before students leave the school setting for employment. The other component that differentiates between placement and cooperative occupational experience programs is the degree to which there is a highly articulated and related classroom instruction program accompanying the cooperative experience program.

School Laboratory

School laboratories, such as school farms, greenhouses, livestock facilities, and forests, are increasingly used to provide experiences needed by students. The increased number of school-owned facilities has somewhat paralleled the difficulty in providing for ownership, placement, or cooperative SOE programs outside the school setting. School laboratories provide many benefits--primarily that all vocational students can participate in SOE programs and that they are in many cases more closely supervised by the teacher since such programs are part of the inschool instructional program. Even though a school program is different from the real work world, many schools are finding it easier to place students in other SOE-type programs after they have had a successful SOE program at a school laboratory. There is a minimum student competency level needed for out-of-school placement, and school laboratories are important in providing the needed minimum competencies.

The major commonality of all SOE programs is that they assist students in acquiring desired occupational competencies. The teacher-coordinator, student, parents, and employer should cooperatively develop a written agreement. This SOE educational plan should list the essential competencies and experiences to be acquired by the student.

Students and employers participating in SOE programs must comply with all state and federal laws pertaining to the employment of youth, including minimum wage and safety regulations. The teacher-coordinator should provide effective coordination and/or supervision of each student's occupational experience. Supervision of occupational experience for self-employed students should be done by the teacher-coordinator. In placement and cooperative SOE-type programs, the teacher coordinates the program, and employers supervise the students.

The teacher-coordinator should have a minimum of one period assigned for SOE planning and at least one period for supervision of students on the job. To determine the students' progress and to assist them in obtaining employment following completion of the program, the teacher-coordinator should maintain a record of supervised occupational experiences for each student.

Each student should keep an accurate and up-to-date financial record of the competencies planned and completed. Students typically receive school credit for supervised occupational experience which takes place during school hours. In some schools, credit is also granted for supervised occupational experience outside the school day and during summers. Through SOE programs teachers of vocational agriculture have an opportunity to individualize education, to get to know the students, and to help students develop the competencies needed for employment.

Future Farmers of America (FFA)

The Future Farmers of America (FFA) operates under a federal charter--Public Law 740 passed by the 81st U.S. Congress. The passage of this law allowed for the incorporation of the FFA as an integral part of the vocational agriculture program for the purposes of developing leadership, citizenship, and cooperation among the students enrolled. This is a unique arrangement which emphasizes that the FFA is a vital part of the education of every student enrolled in vocational agriculture.

Even though the FFA operates as an intracurricular program component, it continues to be recognized nationwide for its contribution to the development of leadership in youth. Leadership--that ability to direct the actions, thoughts, and opinions of others--remains as one of the most important competencies in holding a job. Employers continue to indicate that a lack of personal qualities, that is, affective attitudes and habits, is primarily responsible for people losing their jobs.

Positive job attitudes and habits are needed by all workers. These include:

1. Assuming responsibility.
2. Developing good work habits.
3. Showing initiative.
4. Getting along with others.
5. Being willing to learn a job and to do more than is required.
6. Being flexible, adapting to change, and even searching out innovation.

The FFA is essential for all students enrolled in vocational agriculture, since it provides opportunities for students to develop leadership qualities. Because leadership can be developed, students can acquire those leadership qualities deemed desirable through active participation in the FFA.

The FFA is the leadership laboratory which the students organize and operate. The instructor's challenge as FFA advisor is to see that desirable outcomes result from the learning experiences in the FFA. Some teachers fulfill their advisor roles because students like them and are willing to follow or emulate their actions. Other teachers succeed because they have the ability to influence students by talking to them individually. Still others have the ability to motivate small or large groups through inspiration. Finally, there are those who use power to bring about positive action; they seem to be able to persuade students to do what they want. However, most teachers develop an effective FFA advising style by combining several techniques.

Teacher educators have the responsibility to provide learning experiences that will result in prospective teachers developing the knowledges, skills, and attitudes needed to fulfill their roles as FFA advisors. Teachers will perform their roles best when they know the nature of the FFA organization and the way its various phases can be used to develop leadership qualities in students.

The FFA has become a comprehensive organization. It is comprehensive because it strives to include a variety of activities which provide opportunities to meet the varying interests of students in the seven taxonomies of agriculture/agribusiness. The myriad of FFA activities available tends to overwhelm many students and teachers and may account in part for the recent declining FFA membership in some states. Or, perhaps the decline may mean the organization is not comprehensive enough and/or has failed to meet the needs of a large segment of the population enrolled in vocational agriculture.

The Challenge

American democracy is founded on the concept of shared leadership and wide participation by members of groups in making and carrying out group decisions. This concept permeates almost all groups and organizations and is essential to the successful operation of democratic organization. Leadership qualities are needed by individuals serving their communities or planning recreational activities just as much as they are needed by those operating businesses or running governments. Because of the essential nature of leadership development, it should not be considered extracurricular. Rather, it must be a part of each student's education in vocational agriculture. Membership in the FFA must remain integral, but it must be made so attractive that every student will want to be a part of the FFA and thus will benefit from participation as a member.

Instruction for Out-of-School Youth and Adults

The agricultural education program offered through the public schools need not be limited to persons preparing to enter the world of work, especially when so many adults are in need of additional training or retraining in order to remain productively employed in agriculture/agribusiness. The intent of vocational legislation has always been to work with both inschool and out-of-school persons.

The Declaration of Purpose of the Vocational Education Act of 1968 asserts: ". . . states should improve planning in the use of resources for vocational education and manpower training by involving a wide range of agencies and individuals concerned wih education and training . . . (4.5)." It further states that federal funds authorized for states are to assist them to:

1. Extend, improve, and where necessary maintain existing programs of vocational education.

2. Develop new programs of vocational education.

3. Overcome sex discrimination and sex stereotyping and provide equal educational opportunities to persons of both sexes.

4. Provide part-time employment for youth to continue their education so that persons of all ages in all communities of the state, those who are in high school, those who have completed or discontinued their formal education and are preparing to enter the labor market, those who have already entered the labor market but need to upgrade their skills or learn new ones, those who have special education disabilities, and those who are in post-secondary schools will have ready access to vocational training or retraining which is of high quality, is realistic in the light of actual or anticipated opportunities for gainful employment, and is suited to their needs, interests, and ability to benefit from such training.

Clearly, vocational education is also for adults, and in light of the emphasis on accountability and because of the shorter period for vocational education generally required by adults, it offers quicker results than does vocational education for inschool youth who are preparing for the job market. Most adult instruction centers on updating technology and providing for changing job requirements, new areas of personal interest, or diversified abilities of workers. Because vocational education is funded by the public, it must serve all segments of the population. Clientele who finance the schools through local tax structures must receive consideration. Not only should vocational education programs provide educational opportunities for out-of-school youth and adults for existing jobs, but they should also educate persons who will be able to create employment by opening new businesses in the community.

In production agriculture/agribusiness, the decision to reorganize often results in the employment of additional workers. Also, business and industry are attracted to areas where there are well-trained workers and where schools accept their role of educating out-of-school persons.

Qualities of Successful Adult Educators

Qualities needed by persons who teach out-of-school youth and adults parallel those needed by all teachers. They include:

1. Having appropriate professional training in order to organize, conduct, and evaluate the teaching process.

2. Being technically competent in subject matter areas taught.

3. Being skilled in human relations in order to acquire the resources essential for accomplishing the teaching objectives.

4. Having empathy for students as individuals, in order to stimulate student motivation and meet individual students' unique needs.

5. Enjoying the challenge of helping others meet their individual goals and objectives.

Obligation to Provide Out-of-School Instruction

School policies about adult education and teachers of adults vary widely. Some schools have a policy that all agriculture/agribusiness teachers will be involved in adult education. In others, teachers are hired to teach only adults. Teachers in still other districts instruct both adults and secondary students in varying combinations. Although adult education teaching assignments vary greatly, there is a high degree of esprit de corps among teachers who work with adults. They are at ease with adults in the community and seem more aware of current social and economic problems. They also find professional satisfaction in helping adults solve their problems. Teachers who instruct adults often perceive that they are more readily accepted by adults in the community than are inschool teachers, which results in their having a more positive self-image about their adult relationships.

Experienced teachers of agriculture/agribusiness often comment to beginning adult teachers that they should not fear adult teaching. Rather, they simply need to be open-minded about the process. Through cooperative program-planning processes, adults enjoy helping to identify their needs and to determine competencies to be developed. Adults can also assist in providing needed equipment and facilities as well as helping to bring about the outcomes of the education process. Adult educators often provide an important and needed liaison between the adults in the community and the school system.

As society and technology continue to change, education for out-of-school persons will become even more commonplace--particularly with acceptance of the concept that education is a lifelong process.

Professionalism

Belonging to groups and contributing to the improvement of society through one's work is a goal shared by most individuals. It is this concept which contributes to the growth and development of professionalism among workers. Professional organizations exist in almost every phase of life. Collectively, they exist to help professionals expedite their professional pursuits. A professional organization is of little value if it does not serve the needs of those associated with the profession. On the other hand, how can a professional organization serve its members unless it has the active support of those the organization is to serve?

Prospective teachers of vocational agriculture should leave the university with a clear understanding of the role of their professional organizations and a positive attitude about professionalism. Beginning teachers don't always appreciate the value of belonging to professional organizations. Practicing members can alleviate that difficulty by working with these teachers. One state which consistently has 100 percent of its teachers in their professional organization reports these procedures in working with potential membership.

1. Potential members are introduced at the first professional meeting they attend.

2. Each potential member is assigned a big brother/sister (teacher) to assist him/her in professional development.

3. Potential teachers are initiated into a special teachers organization within the state's vocational agriculture teachers association.

4. Potential members are honored at a special dinner during a conference or meeting.

5. Potential members are visited at their school by experienced teachers in their district.

6. Potential members are able to join their professional organization by post-dating checks until they are on the school's payroll.

7. Every member, as soon as he/she joins the professional organization, is put on a committee and is given duties to perform.

Making members feel welcome, and getting them engaged in the work of the profession as soon as possible, benefits both the members and the professional organization. When new members are actively involved early, the organization usually will not have difficulty in getting them to pay dues and to work for the organization.

The relationship teachers build with their administration and community is especially important. A professional commitment requires much time and effort. Frequently, one must spend time away from the local school and community. It remains a challenge to each teacher to develop and foster relationships with administrators and the community and to point out continuously the benefits derived as a result of being active in one's professional organization. The relationships developed must be mutually beneficial to both the individual and the school system.

A professional organization should not expect its members to contribute money and energy unless the members can see tangible evidence that the organization is accomplishing its goals. In order to accomplish its mission, the leadership of the professional organization must involve its membership in public relations and policy formation activities.

Historically, agricultural educators have achieved some degree of excellence because state associations, which make up the national association, have gained national acclaim. As a result, membership esprit de corps is high, and the benefits to each member are readily identifiable. The professional organizations which serve vocational agriculture teachers are thereby relevant to the membership. To be a professional is to belong to one's professional organization and to participate actively in it.

Public Relations

Public relations (PR) consists of having a quality program and letting others know about it. Public relations is an involved process which keeps the several publics informed about the actions of an organization so that the organization can perform its functions in the social milieu in which it exists. Public relations cannot be performed by one designated person. Every person associated with a particular organization or program plays a vital role, directly or indirectly, in projecting an image to the public.

Wise parents tell their children that they are judged by the company they keep. Likewise, a company, a school, or an organization is judged by its people and the images they leave by their contacts with the public. It then becomes the task of a person or group so designated to relate this concept to its organization and to organize so as to elicit the help of each person in this task.

Thus, a school's image becomes the product of interaction by several groups: local citizens who are served by the schools, teachers who are employed by the citizens, and the administrative body (school board) and school administrative staff who are responsible for the school's policy.

Whether or not these groups are cognizant of public relations and do anything regarding the phenomenon, public relations does exist. The challenge to persons responsible for public relations is for them to present information in an organized way so as to create accurate and favorable images which reach individuals making policy decisions about the local public schools. Long-range policy decisions are often made without accurate information. An organized program of public relations can help ensure against unwise decisions and policies being made by uninformed persons or pressure groups.

Local vocational education teachers should be members of the public relations team. Although the administration most frequently makes the decisions about the program and assigns specific responsibilities, each teacher should try to determine the role public relations should play at the school.

Where there is no overall school public relations program, the teacher of vocational agriculture will often assume the responsibility to plan and carry out a systematic public relations program for the agricultural education program.

Teachers of agriculture have always known that the success of their program is based on the degree to which they are able to gain and maintain public understanding and support. It is paramount that teachers identify their publics immediately. In every public there are persons who influence others. These opinion leaders influence the way others interpret what they read, hear, and feel about the school and the reasons why. These interpretations often determine what is known of the agriculture/agribusiness program. Public relations then becomes a "rifle" rather than a "shotgun" approach in contacting influential opinion leaders. The publics to be reached are many and include both inschool and out-of-school groups. These publics influence policies which govern the local school's agricultural program.

A good public relations program provides for sending and receiving messages about the program of vocational agriculture. This should include relying on personal contact; using the media heavily, particularly radio and TV; and writing a newsletter.

The goal of the public relations program is to create an atmosphere in which the vocational agriculture program can achieve its objectives and develop policies to assure its continued growth and development. The local teacher must play a direct role in the public relations process--it's often the underlying factor that determines the success or failure of the agriculture/agribusiness program.

Summary

The changing role of the vocational agriculture teacher reflects changes in production agriculture/agribusiness. Teacher education in agriculture has a responsibility to update and upgrade its programs to prepare prospective and practicing teachers adequately for this changing role.

CHAPTER 5

Recruiting and Selecting Teachers

William H. Annis
UNIVERSITY OF NEW HAMPSHIRE

Nicholas L. Paul
UNIVERSITY OF NEW HAMPSHIRE

Land-grant institutions have had the major responsibility for training teachers of vocational agriculture since the passage of the Smith-Hughes Act in 1917. Funds from this and subsequent legislation were made available through state departments of education to land-grant institutions for the training of vocational agriculture teachers. The result has been that land-grant institutions have been the primary source of supply for vocational agriculture teachers over the years. In recent times, demand has exceeded supply. The shortage was recognized by the Agricultural Education Division of the American Vocational Association when its executive committee appointed the Professional Personnel Recruitment Committee in July 1965 (5.1). This committee began a longitudinal study of the shortage of teachers of vocational agriculture. The study, published in March 1978 (5.2), represents the thirteenth annual survey of supply and demand for teachers of vocational agriculture. Recognition of the problem has enabled recruitment activities to become a high priority concern of the profession.

One of the basic reasons for the recruitment problem is a lack of knowledge relative to what should be considered in the selection process. A profile needs to be developed which will identify both desirable and undesirable characteristics of teachers of vocational agriculture. The differences between the characteristics identified in the teacher profile can be translated into criteria for the selection process. Recruitment and selection (two distinct processes) can then be directed toward the same goals.

While recognizing that recruitment studies have been limited, Peterson indicates that the research implies that teachers beget teachers (5.3). The identification and recognition of teachers who produce other teachers can provide assistance in recruiting prospective teachers of vocational agriculture. Teacher education departments in the various colleges and universities must develop a selection procedure which will ensure high-level performance by graduates.

Factors Affecting Recruitment

Since World War II, the Commerce and Labor departments' statistics have shown agriculture as a declining industry. This factor alone has caused considerable concern among those working in agricultural education. Since the Vocational Education Act of 1963, agricultural education has, by congressional definition, been "more than farming." Data from these two departments do not include the availability of jobs for students studying vocational agriculture in off-farm areas such as ornamental horticulture, agriculture sales and services, agricultural processing, agricultural mechanics, forestry, and recreation. Recruitment of potential teachers is inhibited by the statistics published by governmental departments. The magnitude of this problem cannot be overemphasized, as students are hesitant to enter a supposedly declining field. Other educators, including counselors, who read these statistics, also advise students to stay away from teacher preparation programs in agricultural education.

The 1976 amendments (5.4) to the vocational act provided for the establishment of the National Occupational Information Coordinating Committee (NOICC), which was intended to correct in part the problem of reporting jobs in agricultural education at the national level. However, states must be careful to ensure the statistics forwarded to the national office reflect the expanded role of agriculture.

Legislation, both federal and state, affects the recruitment of potential agricultural education students. As legislation enlarges the definition of agricultural education, and as more services are demanded of specialists, more job titles involving agricultural education are added. Urban gardening specialists, plantscapers, and small machinery sales and service persons are but a few of the representative job titles in which one can secure entry-level skills through studying vocational agriculture (5.5). The preservice student in agricultural education must prepare for a much wider variety of teaching specialities than was necessary 10 years ago. As this expansion continues, the vocational agriculture profession is experiencing problems in supplying teachers to fill these vacancies. This undersupply of teachers results in recruitment of individuals with non-teaching professional degrees in agriculture in states where a shortage of teachers exists (5.6).

Geographic location has an effect on recruitment programs for preservice agricultural education students. Some 50 to 80 percent of the students who enter colleges of agriculture in land-grant institutions in the Midwest have an agricultural background. In contrast, 80 to 90 percent of the students enrolled in colleges of agriculture in the Northeast and Far West did not grow up on farms or ranches. Generally, students who come from production agriculture areas have a positive attitude toward agriculture, which aids in the recruitment process, for they understand the importance of agriculture as the nation's largest and most important industry.

Internal Factors. The size of the college or university influences the recruitment program. The larger the student body, the greater the possibility of attracting potential students into a preservice teacher education program. Also, the status which agricultural education holds within the college will have a significant bearing on the recruitment of students. Departments which are conducting research and publishing results will be well known throughout the college and will have many potential students referred by colleagues.

At present, there appear to be three ways agricultural education is organized within the land-grant system: (1) within the college of agriculture, (2) within the college of education, and (3) by dual appointments in both colleges. The administrative placement of agricultural education in any of these categories creates both problems and opportunities. For example, if placed within the college of agriculture, the teacher educators find it relatively easy to maintain contact with the agriculture departments. However, such placement means it will be less easy to develop the relationships which should exist with colleagues in the college of education.

There is a tendency to work closely with those individuals to whom one owes administrative allegiance and to have less contact with other colleagues. If housed in the college of agriculture, the tendency is to have less contact with education colleagues. The reverse is true if housed in the school or college of education. There is always the potential difficulty of reporting to two masters when joint appointments are used.

The aforementioned discussions relate to the recruitment problems of agricultural education in that agricultural education must remain visible to draw prospective students. Administrative placement within institutions poses different kinds of challenges for the recruitment program.

Supply and Demand for Teachers of Vocational Agriculture in the United States

A study of the supply and demand for teachers of agriculture in the United States has been conducted since 1965. This longitudinal study was designed by Dr. Ralph J. Woodin of The Ohio State University and was conducted by him until his retirement. Since then, Dr. David G. Craig of the University of Tennessee at Knoxville has conducted this study.

Table 5.1 indicates trends over a 13-year period in terms of supply and demand from the March 1978 study (5.7). The total number who were qualified and the total number who were placed in teaching positions increased fairly consistently during the period. However, the percentage of those qualified who did enter vocational agriculture teaching has remained almost constant at approximately 60 percent.

The summary statistics and recommendations of the 1978 study (5.8) are discussed as follows, with implications for recruitment and selection of vocational agriculture teachers. According to the study, "A total of 1,749 persons were qualified for teaching vocational agriculture in 1977 as compared to 1,038 in 1965." Although the number qualified has decreased somewhat since the high of 1,759 in 1972, the percentage of individuals placed in vocational agriculture has remained steady (see Table 5.1). From 1965 to 1977, the percentage of teacher turnover ranged from 9 to 12 percent. Thus, over the past several years, a teacher turnover of approximately 10 percent has also contributed to the teacher shortage.

TABLE 5.1--Number and percentages of agricultural education graduates entering various occupations

Occupation	Year												
	1965	1966	1967	1968	1969	1970	1971	1972	1973	1974	1975	1976	1977
Teaching vo-ag	64.6	61.4	60.2	61.6	56.9	51.0	49.6	54.8	56.3	58.1	60.2	61.5	60.8
Other work	4.7	8.2	7.2	7.8	7.6	11.0	11.0	11.0	13.7	10.8	9.9	11.0	13.7
Graduate work	9.2	10.0	12.4	7.8	9.3	9.0	9.1	7.9	7.6	8.9	9.8	8.8	6.3
Farming	3.0	2.6	3.3	3.0	3.7	4.9	7.1	7.7	9.3	9.2	8.2	8.2	8.2
Agribusiness	5.6	5.4	3.2	2.0	2.7	4.1	5.1	6.3	6.8	7.8	7.5	6.3	7.4
Other teaching	6.2	5.4	8.2	7.5	11.4	7.3	6.1	6.6	4.1	4.1	3.3	2.5	1.8
Armed forces	6.7	7.0	5.5	10.3	8.4	12.7	12.0	5.0	2.2	1.1	1.1	1.7	1.8
Total number qualififed	1,038	1,151	1,233	1,314	1,566	1,700	1,743	1,759	1,713	1,623	1,660	1,697	1,749
Total number placed in vo-ag	671	701	742	809	891	866	864	964	966	943	999	1,043	1,063

Source: D. G. Craig, A National Study of the Supply and Demand for Teachers . . . (5.7).

A comparison of the total number of vocational agriculture teachers in the nation shows that for the period 1967-1977, the number increased from a low of 10,221 in 1967 to a record high of 12,694 in 1977. In addition, the number of vocational agriculture teachers in technical institutions and community colleges continues to increase. There were 1,509 such positions in 1977.

Several apparent trends reflect the different types of vocational agriculture teaching positions. Recent figures indicate almost 90 percent of all positions occur in general or comprehensive high schools, while about 10 percent of them are in area or vocational high schools. Slightly less than one-half of the positions involve teaching adults and/or out-of-school youth as well as secondary students. The number of teachers in multiple teacher departments represents about 50 percent of the total number, a figure which has stabilized in recent years.

In 1965, compared to 1976, about 230 more teachers were teaching in production agriculture programs and about 170 more were teaching in specialized areas such as agricultural business and supply, ornamental horticulture, and agricultural mechanics.

Also most teaching positions were filled by fully qualified persons holding bachelor's degrees. The number of positions filled by teachers with temporary or emergency certificates increased 1-1/2 percent in 1977.

Recruitment Recommendations

The Professional Personnel Recruitment Committee of AATEA has recommended that approximately 1,800 persons per year nationwide need to be qualified for teaching vocational agriculture. From evidence indicating program growth in new positions, one can conclude that this goal appears realistic. In view of this goal, the following recommendations are proposed:

1. Vocational agriculture teachers should recruit their best students each year to enter agricultural education. Each teacher should have as a goal that at least one of his/her students graduate in agricultural education every three years.

2. Local administrators, state supervisors of agricultural education, and professional organizations should encourage all effective teachers of quality programs to remain in the profession, thus reducing the teacher turnover rate.

3. State supervisors and teacher educators in states with a surplus of agriculture teachers should urge graduates to go to areas where teacher shortages exist.

4. All vocational education groups must continually strive to make teacher salaries competitive with job opportunities in other agricultural fields.

5. State vocational agriculture teacher associations should lead in developing and/or maintaining active recruitment campaigns. They should emphasize the variety of job opportunities--especially

specialized subject areas--the locations of jobs, and the advantages of teaching as a profession--for example, the importance of agriculture and working with youth.

6. Agricultural education leaders at the state level should make strong efforts to reduce the number of uncertified teachers in the profession. They should stress the importance of broadening certification standards to include areas such as ornamental horticulture, agribusiness, agricultural mechanics, and small animal care. They should make sure that the names and addresses of available and certified teachers are placed in the hands of superintendents and boards of education.

7. Agricultural educators should institute a study which will synthesize current research and seek answers to questions about vocational teacher supply and demand. For example, they need to know why qualified graduates do not obtain available positions and why vocational agriculture teachers leave the profession.

8. The Professional Personnel Recruitment Committee should continue its sponsorship of the longitudinal study of the supply and demand for teachers of vocational agriculture.

Recruitments programs have been initiated in response to the increased demand for teachers of vocational agriculture. In 1977, after several years of work, the Standards for Quality Programs in Agricultural/Agribusiness Education was published (5.9). These placement standards have a criteria of 75 percent placement for those students who complete preservice programs.

Recruitment Practices and Activities

Planned recruitment, which is defined as those activities that have the best potential over time of yielding the greatest results is needed. Recruitment in agricultural education, with some exceptions, appears to be conducted without a great deal of planning or concern for cost effectiveness. One must recognize that the recruitment activities selected will determine the growth rate of a program and that certain activities have a greater potential for success than others.

Personal contact with potential agricultural education majors has been rated in the literature as the most effective recruitment activity. According to Lee, "An enthusiastic, healthy and positive attitude toward teaching will be a motivating force in recruiting quality, potential teachers (5.10)." Green conducted a national study to identify, rate, and rank the extent of use and effectiveness of practices which contribute to an individual's decision to become a teacher of vocational agriculture (5.11). Three practices were reported to be of value for the recruitment process. These were (1) encouraging institutions to arrange for agricultural education majors to visit with prospective students at judging contests, conventions, etc., (2) providing college scholarships for agricultural education students, and (3) maintaining active statewide functioning recruitment committees. Recruitment activities utilizing collegiate FFA chapters were also recognized as a practice of considerable importance.

Lee identified a listing of recruitment activities which he called techniques, based on the idea that any strategy or activity that results in a potential teacher actually becoming a teacher can be termed effective. The following is a list of those techniques (5.12).

1. Encourage all professional educators in agriculture to individually contact (by telephone, letter, or in person) potential teachers and discuss the opportunities of teaching as a career.

2. Contact representatives of agricultural businesses or agencies and acquaint them with the need for agriculture/agribusiness teachers. Their help in identifying and recommending potential teachers should also be secured.

3. Make presentations on the career opportunities in teaching to students majoring in other agricultural departments in universities.

4. Use selective mailings to recent graduates or students nearing graduation.

5. Use the news media to develop an awareness of the opportunities in teaching.

6. Develop publications that inform persons about teaching opportunities.

7. Inform potential teachers of the requirements for teaching.

8. Form a committee to coordinate the effort for the entire state.

9. Form a recruitment committee in the professional teachers association.

10. Encourage collegiate groups, such as the FFA, agricultural education societies, or Alpha Tau Alpha, to plan and conduct recruitment activities.

11. Teach a unit in one of the college courses on the teacher's role and responsibilities for the recruitment of potential teachers.

12. Construct displays or exhibits for state FFA conventions, fairs, shows, or other public activities highlighting career opportunities in teaching.

Current Recruitment Activities. In order to identify the type and extent of recruitment and selection activities currently being utilized by teacher education programs in agriculture, a telephone survey was conducted in May 1979 (5.13). Twenty institutions were selected from *A National Study of the Supply and Demand for Teachers of Vocational Agriculture in 1977* (5.14). These 20 institutions were made up of the 10 institutions individually reporting the smallest number of agricultural education graduates and the 10 institutions individually reporting the largest number. The 10 institutions with the smaller number of graduates were identified as having a range of zero to seven graduates. The 10 institutions with the larger number of graduates were identified as having a range of 39 to 92 graduates. Nineteen out of the 20 institutions responded to the telephone survey.

The interview schedule consisted of two basic questions relative to the identity and extent of recruitment activities: (1) What recruitment activities have you used in the past two years? and (2) How many of these activities have you used within the past year? Seven questions were asked relative to selection.

Institutions with Small Numbers of Graduates. Of the 10 institutions selected, 9 were successfully contacted and included in the survey (see Table 5.2). Twenty-one different types of recruitment activities were identified, while the nine institutions carried out a total of 34 recruitment activities.

The activities shown in Table 5.2 are arranged in rank order for the top activities and then by alphabetical order. "Teacher educator visits to schools," the top-ranked activity, was reported by seven of the institutions. The second-ranked activity," staff involvement in FFA activities," was identified by four of the institutions surveyed. Much of that involvement centered around hosting state FFA judging contests on campus. Activities involving the state vocational agriculture teachers association, ranking third, were reported by three institutions. These activities utilized three different practices: (1) a recruitment committee, (2) an article in the association journal, and (3) a scholarship to an agricultural education student. This third-ranked activity was utilized by one-third of the institutions. Two activities were ranked fourth, as reported by two institutions for each activity. Sixteen additional activities, ranking fifth, were reported by one institution surveyed for each activity.

Institutions with Large Numbers of Graduates. The 10 institutions selected were successfully contacted and included in the survey (see Table 5.3). Twenty-one different types of recruitment activities were identified, while the 10 institutions carried out a total of 45 recruitment activities.

The activities shown in Table 5.3 are arranged in rank order. Two activities received the same frequency of use and were reported as ranking first. These were "teacher educator visits to schools" and "vocational agriculture teachers association activities." The latter activities included several different practices: (1) scholarships to agricultural education students, (2) task forces on recruitment, (3) teachers-of-teachers certificates, and (4) journal articles. Both these activities were utilized by 60 percent of the institutions included in the survey.

The next two activities, ranking third, reported by four institutions for each activity were "recruitment brochures," and "staff involvement in FFA activities." Much of the involvement identified in the latter activity centered around hosting state FFA judging contests on campus. The activities ranking fifth were "career days" and "letters to schools." A series of four activities ranked seventh. These were "visits to schools by agricultural education students," "collegiate chapter activities," "newsletters to graduates," and recruitment slides/tapes." All the remaining activities were reported with a frequency of one and are ranked eleventh.

Of these, three of the unique activities reported will be described in greater detail. These include (1) agricultural education orientation courses, (2) funded recruitment project, and program of work on recruitment.

TABLE 5.2--Recruitment activities of the nine teacher education programs reporting the smallest number of agricultural education graduates, 1979

Rank	Frequency	Percentage	Recruitment Activities
1	7	77.8	Teacher educator visits to schools
2	4	44.4	Staff involvement in FFA activities
3	3	33.3	Vocational agriculture teachers association activities
4	2	22.2	Booth at state FFA convention
4	2	22.2	Recruitment brochures
5	1	11.1	Visits to schools by agricultural education students
5	1	11.1	Attendance at district FFA meetings
5	1	11.1	Visits to schools by campus recruiter
5	1	11.1	Career days at high schools
5	1	11.1	Dual-major option
5	1	11.1	Funded recruitment project
5	1	11.1	Interviews
5	1	11.1	High school students invited to campus
5	1	11.1	Letters to FFA alumni
5	1	11.1	Letters to schools
5	1	11.1	Radio spots
5	1	11.1	Recruitment billboards
5	1	11.1	Recruitment slides/tapes
5	1	11.1	Recruitment by student teachers
5	1	11.1	Visits to new and returning teachers
5	1	11.1	Teacher educator visits to high school students

TABLE 5.3--Recruitment activities of the 10 teacher education programs reporting the largest number of agricultural education graduates, 1979

Rank	Frequency	Percentage	Recruitment Activities
1	6	60.0	Teacher educator visits to schools
1	6	60.0	Vocational agriculture teachers association activities
3	4	40.0	Recruitment brochures
3	4	40.0	Staff involvement in FFA activities
5	3	30.0	Career days on campus
5	3	30.0	Letters to schools
7	2	20.0	Visits to schools by agricultural education students
7	2	20.0	Collegiate chapter activities
7	2	20.0	Newsletter to graduates
7	2	20.0	Recruitment slides/tapes
11	1	10.0	Agricultural education orientation courses
11	1	10.0	Booth at state FFA convention
11	1	10.0	Emphasis on holding enrollees
11	1	10.0	Funded recruitment project
11	1	10.0	Guided tours
11	1	10.0	Interviews
11	1	10.0	Letters to FFA alumni
11	1	10.0	Program of work on recruitment
11	1	10.0	Statewide career planning seminar
11	1	10.0	Recruitment by student teachers
11	1	10.0	Universitywide introduction to ag ed courses

Agricultural education orientation courses: This activity is conducted by the Department of Agricultural Education at The Pennsylvania State University. The one-credit course is offered on 15 Commonwealth campuses as a late afternoon course. Master vocational agriculture teachers are hired and appointed to adjunct professor status. This orientation course is the only preservice agricultural education course offered off campus at the Commonwealth campuses, and it results in many students selecting dual majors.

Funded recruitment project: This activity is being conducted through the Department of Agricultural Education at Virginia Polytechnic Institute and State University. The department received a grant for $3,000 to develop a statewide recruitment plan.

Program of work on recruitment: This activity was developed by the Department of Agricultural Education at The Ohio State University. The plan lists the tasks or activities, ways and means, and the individual(s) responsible on a month-by-month basis for September through August.

Combining the recruitment activities of the selected institutions producing both large numbers and small numbers forms the basis for Table 5.4. This table includes 30 different recruitment activities identified by the 19 institutions. A total of 79 activities were carried out. The top-ranked activity, "teacher educator visits to school," was utilized by 13 institutions. Commonality of activities drops rapidly beyond this initial activity. The second-ranked activity was "vocational agriculture teachers association activities." Nine institutions recruited through the professional association. The third-ranked activity identified practices centering on staff involvement in FFA activities. Six institutions utilized recruitment brochures which was rated as the fourth-ranked activity. The fifth-ranked activity, "letters to schools," was used by four institutions. The remaining activities were used by less than 20 percent of the institutions.

Recruitment Activities Summary. Based on the large number of activities identified and the frequency of the activities, it appears that teacher education institutions have identified recruitment as a priority. The numbers also suggest that the recruitment function is largely "just carried out" rather than being planned. Most recruitment activities are not selected based on the numbers of persons recruited as a result of the activity. Recruitment appears to be conducted on an on-going yearly basis without major consideration given to the results of one activity as compared to another. Institutions providing the larger numbers of qualified graduates tend to utilize the state vocational agriculture teachers association more for recruitment purposes, along with recruitment brochures, staff involvement in FFA activities, on-campus career days, and letters to potential students who are still in high school. Institutions providing the smaller numbers of qualified graduates tend to utilize more school visits by teacher educators.

The large number of activities reported suggests that it is not so much the type of activity but the belief that the activity will be successful by the individual or institution conducting it which is important. The data support earlier research findings that successful recruitment depends on the enthusiasm of the individual(s) conducting the activity and that personal contact enhances success.

TABLE 5.4--Recruitment activities of the 19 teacher education programs reporting the largest and smallest number of agricultural education graduates, 1979

Rank	Frequency	Percentage	Recruitment Activities
1	13	68.4	Teacher educator visits to schools
2	9	47.4	Vocational agriculture teachers association activities
3	8	42.1	Staff involvement in FFA activities
4	6	31.6	Recruitment brochures
5	4	21.1	Letters to schools
6	3	15.8	Visits to schools by agricultural education students
6	3	15.8	Booth at state FFA convention
6	3	15.8	Career days on campus
6	3	15.8	Recruitment slides/tapes
10	2	10.6	Collegiate chapter activities
10	2	10.6	Funded recruitment projects
10	2	10.6	Interviews
10	2	10.6	Letters to FFA alumni
10	2	10.6	Newsletter to graduates
10	2	10.6	Student teachers recruit
16	1	5.3	Agricultural education orientation courses
16	1	5.3	Attend district FFA meetings
16	1	5.3	Visits to schools by campus recruiter
16	1	5.3	Career day at high school
16	1	5.3	Dual-major option
16	1	5.3	Emphasis on holding enrollees
16	1	5.3	Guided tours
16	1	5.3	High school students invited to campus
16	1	5.3	Program of work on recruitment
16	1	5.3	Radio spots
16	1	5.3	Recruitment billboards
16	1	5.3	Statewide career planning seminar
16	1	5.3	Universitywide introduction to ag ed courses
16	1	5.3	Visits to new and returning teachers
16	1	5.3	Teacher educator visits to high school students

Selection

Current Selection Activities

The telephone survey conducted in 1979 included seven questions relative to the selection of students once recruited (5.15). These questions ranged from departmental requirements to curriculum organization. The data gathered supported the concept that a department within a college could not require higher admission standards than those of the college. Data from all 19 institutions surveyed supported this concept.

The concept of work experience as an entrance or exit requirement was also addressed (see Table 5.5). Only 1 of the 19 departments of agricultural education reported this as a requirement. Yet, 9 of 19 state departments of education required work experience, ranging from six months to three years, for certification purposes. Thus, work experience in the area of specialization was not found to be a priority selection criterion for departments of agricultural education.

TABLE 5.5--Work experience requirements by agricultural education departments and state departments of education, 1979

Work Experience	Ag. Ed. Depts.			State Depts. of Education		
	Yes	No	Total	Yes	No	Total
Farm/ranch and/or agribusiness	1	18	19	9	10	19

Three of the 19 institutions reported that certification was based on competency rather than traditional courses and credits. Two of these institutions were from the group graduating 39 to 92 qualified vocational agriculture teachers in 1978.

The organization of the institution relative to majors/minors in agricultural education was ascertained to determine if this was related to the number of graduates in agricultural education. This factor was found to be consistent in all institutions (see Table 5.6).

Data on required early initial experience in public schools was also collected. In both groups of institutions, the ratio of a required early

TABLE 5.6--Majors/minors organizational pattern by size of agricultural education graduating class, 1979

Organizational Pattern	Institutions with 39-92		Institutions with 0-7		Total	
	Number	Percent	Number	Percent	Number	Percent
Majors	10	100	9	100	19	100
Minors	2	20	4	44.4	6	31.5
Dual majors	8	80	5	62.5	13	68.4

initial experience in public schools was essentially the same. Only 3 of 10 institutions in the 39 to 92 (large-number) graduates group and 2 of 9 in the 0 to 7 (small-number) graduates group required early initial experience. While the extent and content of this experience varied, it was not found to influence the selection process of one group more than the other.

Efforts were made to determine the effect of equal access legislation on the selection of preservice teachers. Fifty percent of each group indicated this variable had no effect on the selection process. The data are presented in Table 5.7.

Respondents were also asked to specify any other criteria they used in the selection process. The responses varied, ranging from "none" to "student must submit an application," "student must have the recommendation of the agricultural education faculty to participate in the student teaching program," "student must pass communication courses," "student must have ____ grade point average (range 2.0 to 2.5)," "student must be recommended by the department," and "student must undergo a counseling/interview session." All the responses were mentioned by more than one institution.

Selection Activities Summary. The data collected in the 1979 telephone survey indicate a wide variation in the student selection practices used for agricultural education majors/minors at the university level. The selection process begins with the termination of recruitment activities and ends with the placement of certified beginning teachers. However, there are other professional areas which seek to employ agricultural education graduates--professions in which, in some cases, graduates can earn better salaries and in which they can do an equally effective job. As a result, placement in teaching is approximately 60 percent nationwide.

Ideally, the recruitment and selection processes should start at the same time. However, selection should continue after recruitment has been completed. Many teacher educators in agriculture use the personal approach of

TABLE 5.7--Effect of equal access legislation on teacher selection by size of agricultural education graduating class, 1979

Responses	Institutions with 39-92		Institutions with 0-7		Total	
	Number	Percent	Number	Percent	Number	Percent
No effect at departmental level	5	50	4	44.4	9	47.3
Open policy	3	30	3	33.3	6	31.6
Females in classes	7	70	7	77.7	14	73.7
Minorities in classes	5	50	4	44.4	9	47.3
No effort beyond compliance with university regulations	6	60	7	77.7	13	68.4

visiting schools to recruit most of their students. Transfers into the major provide a growth factor in the number of students preparing to teach agriculture.

Summary

The agricultural education profession needs to achieve the AATEA goal of 1,800 persons per year who are qualified to teach agriculture. In order to increase the number of qualified graduates entering the teaching profession, the AATEA 75 percent standard recently projected rather than the current 60 percent level needs to be a priority.

Recruitment must achieve defined success levels. Planning recruitment activities may be the most important factor in defining success levels. Personal contact, as well as an enthusiastic and positive attitude toward teaching, is perhaps the best recruiting activity. Recruitment and selection should begin concurrently, but they must be regarded as separate processes.

The selection activities used do not appear to be based on any clear evidence that relates to successful placement. Rather, selection seems to center around a hodgepodge listing of criteria. Therefore, it is suggested that desirable teacher characteristics be identified which can be translated into criteria for the selection process. The future of teacher education in agriculture must be guided by planned recruitment activities and fostered by desired teacher characteristics in the selection process. Recruitment and selection can then work toward the same goals. These activities are not functions to be carried out just by teacher educators, but rather they should be done by all individuals in the agricultural education profession.

CHAPTER 6

The Curriculum: General Education

James P. Clouse
VIRGINIA POLYTECHNIC INSTITUTE
AND STATE UNIVERSITY

Ronald A. Brown
MISSISSIPPI STATE UNIVERSITY

An important characteristic of a good teacher of agriculture is that he/she be broadly educated, with a strong commitment to and awareness of, the importance of "general education."

Many of the early leaders of agricultural education were broadly educated in the liberal arts, sciences, and areas that extended far beyond the fields of technical agriculture and agricultural education. For a number of years, there was a tendency in many of our teacher education institutions to have rather rigid programs, with a reduced number of electives and greater emphasis on technical agriculture and specifics of agricultural teaching methodology.

Over the past 10 to 15 years, that trend has changed--currently a large number of institutions have more flexibility, particularly in the humanities, social sciences, and liberal arts. As a result, teachers of agriculture today are equipped to work much more effectively with their students and to help students and others in the school community in a more constructive way. Teachers are prepared for an expanded role as community leaders, interacting effectively with many individuals from all spectrums of society.

Concepts of General Education

The term "general education" is sometimes defined as "education of limited use"--the kind which is neither vocational nor academic, that which prepares neither for occupational competence nor for further education. It can also be conceptualized, however, as a necessary component of education rather than as a type of education. Vocational education philosophy has always included general education, even though the inclusion has been more implicit than explicit. Since the Smith-Hughes Act, programs of vocational education in agriculture have been designed to prepare individuals for gainful employment in agriculture. And to varying degrees, the quality of programs at the secondary, post-secondary, and higher education levels has been determined by the success of the persons completing the programs.

Success in agricultural occupations requires both job and general education competencies. General education, in this sense, may be defined as the education that everyone should have, that is, preparation for living effectively (1) with one's self, (2) with one's family, (3) in a community, and (4) as a local, state, national, and world citizen. It can also be viewed as the education that brings to bear the knowledge of society in solving problems. In these areas, the education of all individuals should be fundamentally the same (6.1). This concept of general education may further be described as the phases of learning which should be common to all persons--education gained through dealing with the personal and social problems faced by everyone (6.2). In this context, general education is primarily extrinsic in its value, to be used to modify life rather than to be assimilated for its own sake. It helps individuals to make decisions wisely, to act effectively, to communicate articulately, to think critically, and to adapt well to varying situations in society by identifying and solving problems.

Purposes of General Education

If general education is conceptualized as a useful component of education rather than as a type of education, what are its purposes and how is it useful to teachers of vocational education in agriculture?

A way of estimating the importance of general education in broad terms is to compare the time spent during a lifetime on a job to the time spent away from a job. Another means is to note the causes of success or failure on a job that are related to vocational general education. Studies have shown that a high percentage of employees who are discharged from their jobs have failed for reasons other than job competency (6.3).

Teachers of vocational agriculture need general education. Even today, a significant number of those who prepare for teaching come from small, isolated communities. They may also be from high schools which provide limited general education skills. When these individuals accept jobs as teachers of vocational agriculture/agribusiness, they are charged with developing, implementing, and evaluating education programs. The programs are designed to prepare students for employment in agricultural industry or for further education in agriculture. The four important components of such programs are (1) classroom and laboratory instruction, (2) supervised occupational experience programs, (3) FFA chapters, and (4) adult instruction programs.

Each of the four program components includes tasks which require general education skills. Effective planning of the instruction programs depends upon (1) accurate assessment of the needs of the community, which is necessary for determining the needs of the program; (2) citizen advisory councils, which can aid in the program planning; (3) working with employers, which is essential in developing supervised occupational experience programs for students; (4) broad societal input, which is effective in instituting FFA programs of activities; and (5) good working relationships with agricultural agencies and businesses and with other groups, which can help in planning and conducting instruction programs for adults. Each of these activities requires general education.

Effective teachers must understand, interact with, and often help lead the communities in which they work and live. In various professional publications, authors have spoken of the need for teachers to be involved in programs of general education which contribute to the broad knowledges, attitudes, and skills needed for effective participation in agriculture and in education. Such a program should develop the teacher's ability to communicate effectively; to understand and apply democratic concepts; to think in terms of state, national, and international problems; to participate in civic affairs; to appreciate the cultural and aesthetic aspects of society; to implement the scientific and experimental method; and to work effectively with people (6.4), (6.5), (6.6).

The passing of time makes more obvious the need for teachers of vocational agriculture/agribusiness to interact with educated farmers and agribusiness persons, school administrators, governmental officials, clergy, industrialists, cooperative extension agents, as well as many other individuals and groups. Teachers must be educated for interdependence as well as for independence. Teachers of vocational agriculture/agribusiness cannot be good teachers of students and good advisors to the community unless they are aware of local, state, and national events that are shaping the community's future.

In addition to being prepared for occupational and non-occupational duties, general education teachers can assist in the development of a value system. Values are an integral part of teachers' functioning in and out of their occupational roles; and they govern the teachers' sense of problem identification, prioritization, goal setting, and problem solving.

Changing the General Education Curriculum

Changes in the general education curriculum are continually being made. New and innovative courses are needed and are frequently being proposed. These courses should be taught by the very best professors--those who can command interest, stimulate thinking, and induce action on the part of their students. There has been, and will probably continue to be, more acceptance by students of many general education courses; therefore, it is important that those persons responsible be among the very best professors on a university staff.

University administrators understand the importance of general education and will support a department that takes a positive approach to improving the general education program. Therefore, it is essential that those in agricultural education, regardless of what department they are located in, align themselves with other university units to provide the very best education for their students. It is up to those in agricultural education to insist upon the right to work with general education teachers and to involve them in teaching the undergraduate and graduate students in agricultural education.

Continuing Education

The concept of improved and expanded general education requirements and/or electives may be very difficult to incorporate into undergraduate programs of agricultural education. Thus, it may be appropriate and in the best interest of the total teacher education program to have part of this general education in the graduate program for the professional teacher.

Completion of the master's degree has become commonplace for teachers of agriculture in many states. Others are earning a second professional degree beyond the master's, and an increasing number are obtaining the doctorate. Most states require coursework periodically to maintain teacher certification.

Another factor in continuing education for agriculture teachers is that a rapidly increasing number of the local departments have two or more teachers, resulting in increased opportunities for the release of individual teachers for graduate study. With the area vocational school concept, there has also been a greater demand for specialized teachers, and schools have been more receptive to the idea of permitting teachers to take leave for educational improvement.

It has become accepted in educational circles that the better teacher education programs are not limited to four years but extend to five or six years, and they may be spread out over the work life of the individual teachers. Thus, the general education component of the teacher education program is no longer limited to the four-year undergraduate program. Selected parts of it may now, and in fact should, be incorporated into the continuing graduate program of each and every teacher of agriculture.

Using Available Resources

Agricultural teacher educators continually need to be aware of, and make effective use of, the resources of comprehensive universities. They should be familiar with the facilities of their universities, as well as being cognizant of what types of facilities are available to the staffs and students of other state institutions throughout the country.

If agricultural education students are to know the resources of their institutions for general education, it is essential that the faculty members who advise and direct them be knowledgeable about these resources. There are various ways of acquiring this information. If there is any reward from committee work, it is probably in becoming acquainted with faculty members one would not otherwise come to know. Staff members' research should lead them into contact with relevant sources of information on campus. A department head has the responsibility for knowing and working with the heads of related departments and for establishing communication between his/her staff and the staffs of these departments. A staff member in a department should be responsible for reviewing university publications and communicating with the staff members responsible for those publications which have a bearing upon agricultural education. At many universities, a member of the agricultural education staff is also a staff member in each of the major departments of the college of agriculture. This arrangement provides contact with those persons doing research and preparing publications. Similar relationships with other university departments may be useful.

Some of the university units potentially most helpful may not at first be considered particularly valuable to agricultural educators. Their usefulness may stem from one staff member with a background in vocational agriculture or from a special interest in an area of mutual concern. Sometimes individuals, rather than departments, must be sought out.

As a part of their professional education, majors in agricultural education should get to know the agricultural experiment station and the agricultural extension service of their institutions. Further, the county extension office should be an important early stop for the beginning agriculture teachers.

Increasing Elective Courses

To get the balanced and individualized programs students need may require abandoning rigid requirements and placing more reliance upon student judgment and counseling services. Traditional courses still survive among institutional requirements, and vested interests often keep a strong hold on the teacher education program. Changes that occur are usually made by innovative professors who depart from course outlines. These teachers are few, but they are becoming more prevalent and more vocal.

If more reliance is to be placed upon counseling, it must achieve a higher status than it presently has. Innovative and experienced counselors must be used. Counseling must become more than fitting required courses into a schedule. More must be known about the students who are counseled. Those students who warrant confidence should be given broad options in selecting electives.

Rigid state certificate requirements should, and are, being abandoned to permit university faculties and students to exercise their judgment.

Special Teachers and General Education

Many of the students today are not like the provincial rural students who once majored in agricultural education. They have been living in a different world, have been exposed to television, and have often traveled widely. They know about many occupations other than farming and teaching. They may be in agricultural education courses to get a start in life, with no intention of making a career of teaching. Those who have alternative careers in mind should be allowed to take courses that would help in the exploration of these careers. Courses which help them determine a career choice are a valid part of general education.

Some of the best candidates for teaching are persons who have not decided on teaching as a career and who have not made a final choice of vocation. These may include many of the well-rounded people needed. They are different from those who prepare for teaching agriculture because they admire their high school teachers of agriculture and want to be like them. An early commitment to teaching agriculture seems admirable on the surface, but considering all the alternatives open to youth, it may be symptomatic of a lack of adequate career investigation on the part of a student.

Many teachers of agriculture today are persons who prepared for other careers while in college and then decided to enter teaching and/or came to agricultural education from other types of work. These persons may be even more narrowly specialized than graduates in agricultural education, and they may need considerable general education, which they may be unwilling to get. Some, however, may realize more than undergraduates the limitations in their general training and may be willing to take the needed courses before undertaking teaching.

More and more teachers are preparing for and teaching in fields other than production agriculture. There are differing opinions of specialization among those persons providing agricultural education for these expanded occupations. One group believes that preparation should be narrowly specialized, while another thinks that education programs for these occupations should require twice the time needed to teach a specialty in order to provide time for adequate related courses and general education. Both approaches are probably needed, but it is almost certain the one requiring more time and emphasis on both a strong technical background and a good general and professional education will produce the better teacher.

Agricultural Educators--Proponents of General Education

All agricultural educators have known disillusioned teachers of agriculture who do not recommend teaching to their students. Only a relatively small percentage of teachers regularly have graduates who prepare for teaching. These teachers' disillusionment may be traced to their having been prepared narrowly for one career, which has not been satisfying to them but from which they have been unable to escape. Does the profession want in teaching those who go into it because they do not know the alternatives? The university must provide each of its students with a broad view of the world of work and the myriad of career opportunities available. The profession is obligated to permit and encourage students to view occupational opportunities widely. It must keep them from becoming victims of a narrow and provincial education.

Good teachers of agriculture have always recognized that a lack of knowledge in basic subjects must be remedied in agriculture classes. Often they have helped students see the usefulness of the other subjects they are taking and have made these subjects relevant by applying them to agricultural situations.

Many teachers of agriculture have improved their ability to recognize and teach students with special social, economic, physical, or mental needs. Frequently these students are deficient in general subject areas, and teachers of agriculture should give them the necessary remedial help, or they should arrange for such help. Agricultural educators at the university level must continue to influence and encourage local teachers of agriculture to improve the general education of this nation's rural and agriculturally based people.

General Education for Leaders in Agricultural Education

We are living in an age of specialization. Even the field of agriculture is too broad to be one specialty area. As we develop more multiple-teacher departments and more area schools, the demand for specialization will increase. No one can be an expert in the entire field of agriculture. Narrowing the area of specialization affords opportunities for broadening general education. Agricultural educators should not react by requiring that their students merely know more and more about less and less.

The challenge to provide more and better general education is particularly strong in the preparation of the future leadership in agricultural education. The current leadership will hopefully be sufficiently aware of its limitations so that it will ensure that the next generation of leaders is not similarly handicapped. Given a broadly educated and experienced leadership, the other changes needed in our programs can be made.

There is need for leaders who are able to deal with all kinds of people. The clientele of vocational agriculture are no longer only farmers. In preparing individuals for agricultural business and industry, agricultural teachers meet not only the operators of small, independent establishments but also the executives of national corporations. Whether the education needed is labeled as vocational or general, agricultural educators must become familiar with all areas of agriculture. The new territory agricultural educators are entering is far larger that the territory traditionally occupied. In it agriculture teachers encounter big business and industry, powerful labor, and governmental bureaucracy. Restrictions upon program operations not previously known now exist. Even if future teachers are to work in rural farming communities, they must be aware of the broad relationships of the agriculture in these communities.

With a far-reaching vision of expanding relationships, it is possible to see opportunities for agricultural education never before foreseen. Too many arbitrary lines have been drawn; a broader perspective is needed. There should be an awareness that every human being needs some kind of agricultural education. Agricultural educators need the kind of general education that will cause them to "see life steadily and see it whole."

Most of all, the profession needs education which can give it a balanced and justifiable set of values by which it can judge alternative courses of action.

Agricultural education students should not be mislead by false values or values of secondary importance. They should not be excluded from programs which stress values not emphasized in agricultural education.

The finest product that could be produced would be a person committed to high values after exposure to all the alternatives, one who has the knowledge and the techniques required for implementing these values. The preparation of this type of person is within the possibility of our teacher education system. The result will come from the right combination of technical, professional, and general education tailored to individual and programmatic needs.

General Education as a Part of Professional Teacher Education

The National Council for Accreditation of Teacher Education continues to emphasize the importance of general education by stating in Standard 2.2 that "there is a planned general studies component requiring that at least one-third of each curriculum for prospective teachers consist of studies in the symbolics of information, natural and behavioral sciences, and humanities (6.7)." The council further emphasizes this need for a general studies component by suggesting that the humanistic and behavioral studies participated in by teachers should differ from the usual study of such subjects and should emphasize and address themselves to the problems of education. These humanistic and behavioral studies as explored in education require a familiarity with the parent disciplines on which they are based. Thus, a strong general education component is an important part of any teacher preparation program.

Other groups, including the National Academy of Sciences, have historically recommended that as much as one-half of the undergraduate program in agriculture be devoted to general education. In all cases the accrediting agencies, learned societies, and other regulating and recommending groups consider general education to be education intended to support or undergird a person's professional speciality.

Over the past 20 years, there has been a general trend to reduce the number of credit-hours of agriculture required for graduation in agricultural education. Some of this has resulted from a general reduction in the total number of hours required, but at a number of institutions the reduction in agricultural courses has resulted in an increase in the general education requirements and in the number of free electives. As a result, there has been a greater opportunity for the student, through proper advising, to elect more courses in literature, art, music, political and social science, sociology, economics, and other liberal arts subjects.

General Education in Professional Courses

Much general education continues to be a part of those courses taught by agricultural teacher educators. Courses in professional education for all secondary and/or elementary teachers contain a great deal of content that is genuinely general education. Similarly, many teacher education courses offered and taught by agricultural educators are taken by a variety of vocational education undergraduates, for example, those in agriculture, home economics, business, and distributive education. Much education of a general nature is frequently provided through these courses.

Agricultural education students can profit from courses in educational psychology, educational sociology, educational history and philosophy, and comparative education. A comprehensive college of education has a diverse faculty who are trained in a multiplicity of areas, with depth in economics, sociology, psychology, history, political science, and philosophy. As a result, some teacher educators are as effective in teaching these subjects to their agricultural education students as their colleagues are in the subject matter departments.

It is difficult to draw a line between general and professional education. Agricultural educators need to understand community power structures, personal and class prejudices, ethnic and religious differences, historical backgrounds of current situations and events, organizations of many types, demographic trends and their implications, governmental structure and public policies and the involved processes of making public policy. Knowledge in these and many other areas will be of value to students as teachers and as citizens.

Training in the natural sciences has long been a basic and vital part of agriculture. However, the social sciences related to agriculture have been an important latecomer and, in some cases, have had some difficulty surviving in agricultural curricula. Agricultural education students get their work in the natural sciences; the struggle is for them to get the general education needed in the other fields.

In addition to the coursework, there is always an opportunity for students to acquire some general education cognitive and affective learning through extracurricular college or university activities. The agricultural education organizations such as Alpha Tau Alpha, collegiate FFA, or an agricultural education society or club can supply some of the general education elements missing in the agricultural education curriculum. Student activities in fraternities and residence halls and at lectures, musical events, and campus-related churches provide others. Agricultural educators can do much to stimulate student interest in these activities.

General Education Through Teaching Methods

An excellent source of useful general education is the teaching methods used, regardless of the subject matter taught. Students can be taught how to "think" in any course. Like most college students, agricultural education students know how to choose professors, rather than courses, and when given the opportunity, they invariably select instructors with experience and a depth of understanding of both people and subject matter. Students can learn methods of study and research in any field. Thus, each student should be encouraged to give major attention to one or two agricultural areas and to learn well the process of dealing in depth with that or those areas. The process is then readily transferable to other areas.

General Education for Agricultural Leaders

Agricultural teacher educators are often preoccupied with preparing teachers to do well on their first jobs. This practice is increased by curriculum requirements from colleges and universities, state departments of education, state governments, and professional associations. When all these requirements are met, nearly all the credits needed for graduation have been completed. In such a situation, a minimum of general education courses is required, and students are left with little opportunity or encouragement to elect other general courses. When they exercise their limited opportunities for election, they commonly choose more courses in their fields of specialization.

Teacher education programs which allow or encourage a minimum of general courses are usually focusing on teachers doing well on their first jobs, which is important. However, to ignore that teachers have the right to climb the career ladder is ethically wrong--and to fail to help potential teachers plan for this mobility is unjust. Many teachers, because of, or in spite of, their education, leave the teaching field and become farmers, employees of various agricultural agencies, lawyers, bankers, and politicians, as well as education leaders in various capacities. Good leadership by individuals who understand agriculture and education is essential in these occupations and all segments of society. Teachers of vocational agriculture/agribusiness need to be prepared for such career mobility. If not, who will occupy the roles of high school principal, county superintendent of education, school board member, community college agriculture teacher, state department of education supervisor/consultant, agricultural teacher educator, and vocational director?

This concept of preparing teachers for career mobility may appear to be juxtaposed to preparing teachers for success on their first jobs. This, however, is a misconception. The same skills and abilities needed for career mobility are needed by teachers for success in their first teaching assignments. Teachers need to be able to deal with all kinds of people. In preparing persons for the agricultural industry of today and tomorrow, agriculture teachers work with a wide range of people. They confront not only the operators of small establishments but also the executives of large corporations. The current territory of agricultural educators is two to three times as large as that of the past. Agriculture/agribusiness teachers now work with big business and industry, powerful labor, and bureaucratic government--and they must be equipped to cope effectively with each.

This is a time when specialization is required; even the field of agriculture is too broad a specialty. With the advent of more multiple-teacher departments and with more area schools teaching agriculture, the demand for specialization is increasing. Therefore, it is no longer necessary for an individual to try to become an expert in the entire field of agriculture. Narrowing the area of specialization affords increased opportunities for broadening general education--if agricultural educators do not react by requiring that their students know more about a specialized area.

As teacher education curricula are revised to meet new needs for knowledges and skills and abilities in agriculture, curricula should also be adjusted to better meet the needs of future teachers to prepare themselves not only for an occupation upon graduation but also for "life," which may include several occupations and a host of other responsibilities.

Teacher education departments also have an obligation to help those students who may not choose to teach after graduation. Approximately 50 percent of the students who graduate from teacher education programs do not accept teaching positions as their first occupation. This, however, does not release teacher educators from an ethical responsibility to those students. Many of these students, based on good vocational and general preparation, can move directly into related occupations such as those in the Cooperative Extension Service, Farmers Home Administration, Federal Land Banks, and Soil Conservation Service. Some teacher education departments do offer an educator curriculum option to students who do not wish to be certified as vocational agriculture/agribusiness teachers. In many colleges and universities, teacher education departments meet the needs of these students more effectively than other departments can. Individuals who make this choice also need general education courses and activities included in their education programs.

General Education for Living

In any kind of endeavor, the goals or objectives must be clearly thought out and stated. This is also true of general education. What are the goals and standards that are to be attained and what steps need to be taken to reach those goals and standards? If these are clearly spelled out and taken, then much of the confusion and intangibility that have been associated with general education disappears.

National goals for general education date back to some of the early efforts in this country in providing quality education for young people. One of the early influential efforts concerning goal development was the Seven Cardinal Principles, developed in 1918. At least three of these were directly concerned with general education. More recently, there have been statements about the imperative needs of youth and goal statements concerning the purposes of education in American democracy. In 1955, the White House Conference on Education developed 14 basic goals of education. And, in 1961, the National Education Association stressed the importance of the educational system teaching students the "ability to think." One of the most recent efforts to develop broad goals of education took place in 1973, when 18 statements were developed through the efforts of Phi Delta Kappa (6.8).

In all these goal or purpose statements, there is continued emphasis on general education and the importance of general education as being fundamental for living.

There are always life activities in addition to the vocational activities participated in by most individuals. Teachers, like other people, marry and rear families, interact with others socially, participate in organizations, vote to elect public officials, and engage in leisure activities. These "living" activities may be performed ably or not, just as any vocational activity is performed. Many who are competent vocationally fail in the other areas of living. The age-old question concerning whether it is more important to make a living or to make a life keeps reoccurring. This question is fundamental in providing the best general education.

General Education: Luxury or Necessity?

General education is not something that we can take or leave. It is an indispensable ingredient in preparing teachers in agriculture. As such, it should be planned, implemented, and evaluated just as seriously as any other part of the teacher education program.

As teacher education curricula become more specialized, they should also be improved in terms of general education courses. A school principal who only wants a technician should not be our focus--rather, our focus should be on helping individuals prepare for life in a complex society, with the tangible product being persons who can help students become successful employees and responsible citizens.

Summary

General education is an integral and important part of teacher education in agriculture. It helps students to (1) develop basic skills and values, (2) identify and solve problems, (3) set goals, and (4) become community leaders. Agricultural teacher educators need to be prepared to help students adapt to life activities after graduation--to achieve success and satisfaction in agricultural employment and to become responsible citizens.

The teacher education curriculum should be a mixture of technical, professional, and general education tailored to individual and programmatic needs. Thus, general education should be planned, implemented, and evaluated just as rigorously as any other part of the teacher education curriculum.

CHAPTER 7

The Curriculum: Agricultural Subject Matter and Occupational Experience

J. David McCracken
THE OHIO STATE UNIVERSITY

> A long founded principle widely held for teachers of vocational agriculture is that the teacher should be able to operate from the perspective of the real world. This means that the curriculum in our colleges of agriculture must provide practical, hands-on experience to develop skills as well as knowledge. Principles should be taught and understood, but they can best be understood if the student has an opportunity to engage in application of the principles (7.1).

The preparation of vocational agriculture teachers in their subject matter fields has been of concern to agricultural education throughout its historical development. Many writings early in this century emphasized the need for teachers to be competent in agriculture. Related to this concept is the insistence of early agricultural teacher educators upon actual farm experience as a prerequisite to entrance into the agricultural teaching profession. In more recent years, many institutions have recommended prospective teachers gain this experience in agricultural occupations directly related to their teaching specialties. A concurrent expectation is that only colleges in a position to offer an indepth curriculum in agriculture should prepare teachers of agriculture. Qualifying institutions provide adequate faculty, library, and laboratory resources.

Why were the early leaders in agricultural education so concerned with the technical competence of vocational agriculture teachers? "The job of the teacher is to teach. It follows, then, that the teacher must have a degree of mastery of the subject matter he teaches (7.2)." This mastery must encompass both the knowledge and skill of the discipline. Teachers who can interpret and apply basic principles and demonstrate practical skills needed by students will bring credibility to themselves and their profession.

The importance of agricultural subject matter in the curriculum was recognized by those establishing standards for high quality programs in agricultural education. An adopted teacher education standard states (7.3):

> The agricultural education curriculum provides for adequate preparation for teachers of vocational agriculture/agribusiness with course work distributed as follows:
>
> General Education 20-30%
> Professional Education 20-30%
> Technical Agriculture/
> Agribusiness 30-40%
> Electives 10%

Clearly, those drafting and approving this standard believed that agricultural subject matter should receive greater emphasis than other parts of the curriculum.

In designing undergraduate curricula to prepare vocational agriculture teachers, teacher educators and others must be able to answer some fundamental questions. These include:

- What trends are influencing the technical competence of prospective teachers?
- What influence will the performance-based teacher education movement have on the undergraduate teacher preparation program?
- Should the curriculum in agricultural education include broad and comprehensive agricultural training, or should it include specialization in one or more subject areas?
- What is the proper balance between the teaching of principles and the demonstration of skills and techniques?
- What are the alternative methods of developing technical competence?
- What institutional support is needed to aid students of agricultural subject matter?
- What criteria may be used to certify that students are competent in the skills and knowledges they will be expected to teach?
- Who is responsible for developing the agricultural competence of prospective teachers?

The issues and concerns raised by these questions must be addressed by the agricultural education profession.

Program planning, teaching, and student advising in teacher education in agriculture should prepare future teachers to enter vocational agriculture teaching committed to technical excellence both for their students and for themselves.

Trends Influencing Technical Competence

Numerous changes have taken place since the time when early leaders in agricultural education insisted that persons entering college to prepare for teaching be farm boys--preferably farm boys who had taken vocational agriculture in high school. The demographic background of students has changed. The agricultural curriculum at all levels is different, as are the institutions preparing teachers. Teacher certification requirements are changing. Budget restrictions in institutions of higher education and limited financial support for teacher education are bringing pressures which affect programs of teacher education in agriculture.

Undergraduate Students

An increasing percentage of undergraduate students in agriculture have a non-farm background and little, if any, practical experience in any phase of agriculture. In many colleges, only about half the students have agricultural backgrounds. Furthermore, the trend is toward a smaller percentage of students in the undergraduate agricultural education curriculum who have had vocational agriculture in high school. These trends may be because there are fewer farms in existence, there is less emphasis on college enrollment for students in vocational agriculture occupational preparation programs, or other factors. As a result, the lack of student background experience in agriculture must be taken into account by teacher educators. More of the experiences formerly considered prerequisite must now be provided within the preservice curriculum.

Specialization of Agriculture

Since the enactment of the Vocational Education Act of 1963, vocational agriculture has encompassed much more than production agriculture. New programs have been established in agricultural supplies and services, agricultural mechanics, agricultural products, horticulture, renewable agricultural resources, forestry, small animal care, and horse management. But, even more dramatic than the changes in the programs of vocational agriculture has been the change in agriculture itself. There are fewer general crop and livestock farmers each year, as more farmers are specializing in swine, corn and soybeans, or dairy cattle, for example. In agricultural industries, job descriptions are becoming more specifically defined. Thus, there is a decreasing commonality in the technical competencies required to teach in the different instructional areas in agriculture. The technical preparation of a floriculture teacher must be different from the technical preparation of an agricultural mechanics teacher. Therefore, prospective teachers must develop specialized areas of expertise if they are to be competent in preparing students to meet the needs of the changing agricultural industry.

Undergraduate Curricula

There is pressure within many universities to reduce the number of credit-hours taken by prospective teachers in technical subject areas. Some institutions have reduced the number of quarter credit-hours required for graduation from 196 to 180 (semester credit-hours from 130 to 120). Fewer total credit-hours often means fewer credit-hours in agricultural subject areas. Because students often do not enter college with prerequisite occupational experience in the agricultural specialties in which they intend to teach, internship experiences are provided as a part of the undergraduate curricula. Such internship experiences are needed, but additional "credit for work experience" may result in fewer credit-hours in agricultural subject matter courses. This means that students now obtain credit for experiences formerly considered as prerequisite for matriculation.

The preparation of students in general education, professional education, and technical agriculture results in discussions concerning the proper balance between the three areas. If teacher educators determine a greater share of the undergraduate curriculum should be devoted to professional education to prepare professionally competent teachers, they should keep in mind that this also will be reducing the portion of the curriculum that can be devoted to preparation in technical agriculture. The impact of additional courses in professional preparation and additional field experience and student teaching credit-hours must be carefully weighed. Related to this concern is the possibility of universities' requiring additional emphasis in general studies to maintain credit-hours taught in these areas in the face of declining enrollments. Teacher educators in agriculture should support a reasonable allocation of student credit-hours to general studies, but only if such support does not result in less competent teachers of vocational agriculture.

Institutions Preparing Teachers

The number of universities preparing teachers has increased in recent years.

> Due to the shortage of teachers of agriculture, the clamor for more students by some institutions and the availability of some vocational education funds, institutions have attempted to prepare teachers of agriculture without sufficient resources. Standards need to be developed and followed in this regard (7.4).

Bender notes that states which have more than one teacher preparation institution are having just as much difficulty in securing a sufficient number of teachers as those states which have only one such institution (7.5). The breadth and depth of agriculture programs that are available should be a fundamental consideration in approving university programs for teacher education in agriculture.

The Standards for Quality Programs . . . for teacher education states (7.6):

A minimum of four (4) FTE faculty are employed to meet the technical education requirements of students in each of the following areas:

a. Agricultural engineering and mechanics
b. Plant and soil science
c. Animal science
d. Agricultural economics and business management

Comparable FTE faculty are assigned in specialized areas of certification (e.g.: ornamental horticulture, agricultural products processing, forestry, and natural resources).

Teacher educators will have difficulty offering the high quality technical preparation needed in specialty areas in institutions which are not committed to excellence in agriculture.

Teacher Supply and Standards

A pervasive and continuing problem of the profession has been the shortage of vocational agriculture teachers. The numbers of teachers qualified has remained fairly static from year to year, but the demand has risen steadily, especially in non-farm teaching areas. This shortage has resulted in various "temporary" or "emergency" certification procedures to qualify personnel as teachers of agriculture.

> Most of these teachers have been graduates from other departments in the colleges of agriculture without the regular professional preparation for teaching. In some states prospective teachers have been recruited from business and industry on the basis of the experience they had in the taxonomy area in which they needed to teach regardless of their college preparation (7.7).

When teachers are recruited directly into teaching without college preparation, it is difficult to establish the breadth and depth of their technical competence. Years of work experience may be a poor indicator of technical competence. The background of the individual recruited in this manner may result in overemphasis on skill development as opposed to the teaching of principles and the development of managerial abilities. Teacher educators must assume the responsibility for improving the technical as well as the professional competence of those entering teaching through alternative routes.

Furthermore, the shortage of teachers may encourage some states to consider overlooking the enforcement of occupational experience as a requirement for certification. The teacher can then be expected to have limited credibility with the agriculture employers who are the source of jobs for that teacher's students. The shortage of teachers is a continuing problem, one that may not be best resolved by allowing those individuals not possessing the necessary technical competence to enter the profession.

Budget Restrictions

Competition is increasing for funds needed to support higher education. Universities faced with fewer real dollars to operate programs may fail to fill vacant faculty positions, replace outdated equipment, enlarge laboratory facilities, or initiate new programs that are needed. As vocational education in agriculture has expanded and become more comprehensive in scope, university programs have been faced with budget reductions. Teachers in secondary and two-year post-secondary institutions often teach in facilities which are much better than those found in teacher education institutions.

The Standards for teacher education states: "Modern livestock, greenhouse, agricultural mechanics, and experimental farm facilities are used in the teacher education program (7.8)."

University administrators may choose to deal with the budget crunch by offering fewer laboratory or skill courses, emphasizing theory apart from practice. Another temptation is to offer larger sections of courses, making the development of skills more difficult. College students will find it difficult to develop needed agricultural skills in a university faced with constant financial restrictions.

Competency-based Teacher Education

There is growing recognition that competency-based teacher education must be concerned with technical as well as professional competence. Much research relating to the development of professional competence has been accomplished. A major effort in which teacher educators are identifying technical as well as professional competencies is at Cornell University. The competencies are being verified by specialists from technical departments in the college of agriculture and life sciences, by agricultural industry specialists, and by experienced teachers.

A major recommendation stemming from a study conducted at The Ohio State University is that the resulting task inventories should be used as a resource in developing the technical agriculture competencies needed for competency-based teacher education programs (7.9). There are few occupational areas in agriculture in which there is not at least some kind of competency study. Recently, approximately 50 persons from 40 states participated in a national project (7.10) to develop job descriptions and competency lists for approximately 200 agricultural occupations. Teacher educators need to make use of existing information in providing opportunities for students to develop competence in knowledges and skills needed for teaching vocational agriculture. Students preparing to teach get much of their coursework from the various subject matter departments in colleges of agriculture. Because these courses are offered by departments as a "service," teacher educators may have little influence concerning course content. Teacher educators must tactfully communicate the needs of their students to these subject matter departments; but at the same time, they must keep in mind that the true specialists in the technical agriculture areas are in these departments.

Defining Technical Expertise

Decisions concerning the breadth and depth of technical competence needed by teachers are not easily made.

> Is it really necessary for a teacher to possess <u>all</u> the competencies on the production agriculture list in order to be a qualified teacher? Many teacher educators would readily admit that they were not competent in all areas of production agriculture at the beginning of their first year of teaching. Some may have been very competent in some areas and somewhat deficient in others. They were able to succeed by involving community resource persons, studying with the students, additional formal study, and concentrated practice. Interesting questions for further research remain. What are the levels of mastery needed to succeed in teaching? What is a desirable breadth of technical competence to possess for acceptable performance in teaching (7.11)?

Technical and/or Professional Competence

> Should we even attempt to establish minimum levels of technical competence? It is possible that one teacher might succeed with little technical competence and another fail even though the level of technical competence is high. It is for this reason that a hypothesis is forwarded: minimum levels of professional and technical competence required for success in teaching are inversely proportional. In other words, when the level of technical competence is high, a teacher can succeed with less professional competence than normal. When the level of professional competence is high, a teacher can succeed with less technical competence than normal (7.12).

Breadth and Depth of Subject Matter

As a group, teachers of vocational agriculture are becoming increasingly heterogeneous. Some may teach in only one specialized area in agriculture, while others may teach in several areas. Some may teach just career orientation programs in grades 8 through 10, while others are involved only in occupational preparation programs. Those teaching in occupational preparation programs may be working at the secondary, post-secondary, and/or adult level. It is likely that the breadth and depth of technical competence significantly varies from teacher to teacher. The teacher education program must be designed to prepare students for the teaching role(s) they are likely to enter. Inservice programs must of necessity be used to continue the development of teachers in subject matter fields. It is becoming increasingly important, however, to prepare specialists who also have a general knowledge of agriculture. This is a different concept from the past practice of preparing generalists who might also have emphasized one or more specific areas in their undergraduate programs.

Principles and Underlying Theory or Skills

Educators who endorse the competency-based approach to establishing technical competence should use existing competency lists with caution.

> Existing competency lists are designed to communicate the tasks essential for occupational success. They are written in terms of what the worker does. Should teachers be prepared in only these "doing" skills? To what extent must beginning teachers also understand the principles and underlying theory supporting the performance of job skills (7.13)?

In reality, both the principles and underlying theory and the mastery of skills and techniques are important.

> The problems of agriculture are rooted in adjustment to changing conditions, most of them economic and caused by changes in technology. Increasingly, proficiency in modern agriculture requires more mental acumen than manual skill. Thus the ability to interpret and apply a basic principle becomes more strategic than the ability to execute a specific skill. . . . Indeed the application of a principle depends on mastery of skills and techniques. . . . A principle is only of academic value until it is applied (7.14).

Developing Technical Competence

Students use alternative routes to gain the necessary technical competence to teach vocational agriculture. College subject matter courses are a major means of developing students' technical competence. Such courses are usually offered in subject matter departments in four-year universities with agriculture colleges. Some students enroll in an agricultural curriculum in a community college or a technical school for one or two years and then transfer to a four-year university to complete the requirements to enter teaching. Most teachers upon graduation continue to enroll in agricultural coursework on an inservice education or graduate education basis. Another major means of developing agricultural competence is the use of occupational experience programs for prospective and practicing teachers. These internships, which involve placement and supervision by a university faculty member, are specifically designed to develop needed skills and to provide students with real work experience in settings similar to those for which they will be preparing their students. Such experiences, if properly planned and supervised, can also be used to teach prospective teachers how to manage placement experiences for their students.

Many students have previously acquired needed agricultural skills through work experience, in coursework at the secondary level, or through programs in technical schools. Because of this, some universities provide credit for students' former work experience. Emergency or temporary certification programs often encourage the hiring of prospective teachers who are deemed to be technically competent. A sequence of professional preparation courses is then provided to complete certification requirements. The level of technical competence is measured by years of experience or by occupational competency testing.

College Curricula

The trend towards increasing specialization in agricultural education is influencing the way undergraduate teacher preparation curricula are designed. In a 1976 survey of 17 teacher education institutions in the Central Region, 9 institutions offered certification in at least one specialized area (7.15). Differing options are being offered to students, depending upon the specialized areas for which students desire to certify. Students preparing for post-secondary or adult instruction may also need programs different from the ones offered to teachers preparing for secondary instruction.

Table 7.1 reports the percentage of institutions in 1953 (7.16), 1965 (7.17), and 1976 (7.18) by the hours of agricultural subject matter required for the B.S. degree in agricultural education. The trend is clearly toward requiring fewer hours. The medial requirement in 1953 was in the 55- to 59-semester-hour category; in 1965, it was in the 50- to 54-hour category; and in 1976, it was in the 46- to 49-hour category. It should be recognized that some institutions require more related science courses to counterbalance the lower number of required hours in agricultural subject matter. Students often choose to take additional agricultural subject matter courses as electives. In the 1976 survey, the hours of free electives available in the various undergraduate curricula ranged from 0 to 30, with a median of 15. The technical agriculture requirements reported in the 1976 survey appear to fall within the guidelines of the national standards. Teacher education Standard 13 recommended 30 to 40 percent of the undergraduate curriculum be devoted to technical agriculture (7.19). Current practice seems to be within the 35 to 40 percent range.

There is increasing pressure on the undergraduate curriculum, partially caused by fewer credit-hours required for graduation. Examination of Table 7.2 reveals the median semester-hours required for graduation were about 135 in 1953, in the 136- to 140-hour range in 1965, and in the 126- to 130-hour range in 1976 (7.20), (7.21), (7.22).

At The Ohio State University, students must have a minimum of 40 semester-hours in agriculture, of which 16 must be in the specialized area in which they will be certified to teach. Most students elect to take additional hours in agricultural subject matter. The Ohio State University and many other institutions are offering dual-major programs. A student may major in agricultural education and also meet the requirements for a major in another department within the college of agriculture. Dual-major programs are available with the departments of agricultural mechanization and systems, horticulture, agricultural economics, animal science, dairy science, poultry science, agronomy, and natural resources.

TABLE 7.1--Semester-hours of agricultural subject matter required for the B.S. degree in agricultural education

Hours Required	Percent of Institutions		
	1953	1965	1976
	N = 48	N = 30	N = 17
Under 40	2	20	18
40-45	10	13	18
46-49	10	17	18
50-54	23	23	35
55-59	19	17	12
60 and over	35	10	0
Total	99	100	101

Sources: C. Oscar Loreen, "A Study of the Agricultural Education Curricula . . ." (7.16); M. J. Peterson and A. P. Torrence, "The Curriculum: Agricultural Subject Matter" (7.2); and J. D. McCracken, Status of Pre-service Teacher Education in Agriculture, Central Region (7.15).

Occupational Experience Internships

As a group, students entering preservice programs in agricultural education are more diverse in terms of background and experience than was true in earlier years. The proportion of female students aspiring to become vocational agriculture teachers is increasing steadily. The number of students with non-farm backgrounds also continues to grow. Even students who enter with farm backgrounds do so without the breadth and depth of agricultural experience of former students. Many possess experience in limited areas only, which reflects agricultural specialization. Many students preparing to teach vocational agriculture are seeking help from teacher educators to secure the needed practical occupational experience in technical agriculture.

Certification criteria in many states suggest a need for teacher preparation activities that assist in developing occupational competence, that is, skills in performing various jobs and knowledge of various job-related behaviors (7.23). In the 1976 survey of 17 institutions in the Central Region, the months of occupational experience required for certification varied fron 0 to 36. The median was 12 months. Meeting the requirements for a period of occupational experience has often been a responsibility of the students to complete, without much help from the universities.

TABLE 7.2--Total semester-hours required for the B.S. degree

Hours Required	Percent of Institutions		
	1953	1965	1976
	N = 48	N = 30	N = 17
Under 126	10	7	29
126-130	29	13	53
131-135	12	13	12
136-140	26	43	6
141-145	19	17	0
146 and over	5	7	0
Total	101	100	100

Sources: C. Oscar Loreen, "A Study of the Agricultural Education Curricula . . ." (7.16); M. J. Peterson and A. P. Torrence, "The Curriculum: Agricultural Subject Matter" (7.2); and J. D. McCracken, *Status of Pre-service Teacher Education in Agriculture, Central Region* (7.15).

Several developments have provided the basis for the development of occupational work experience activities as an integral component of the preservice and inservice program in teacher education. First, the rapidity of technological change in the agricultural sector often makes previously acquired occupational skills obsolete within a short period of time. Therefore, there exists a demand for opportunities by which undergraduate students and currently employed teachers may up-date and/or develop technical agricultural skills (7.24).

• • •

The limited work experience of many undergraduate students and teachers provides a second reason for the incorporation of practical work experience as a component of the preservice and inservice program. . . . many of today's undergraduate students aspiring to become vocational horticulture teachers do not have parents who owned and operated a family horticulture business. A majority of the undergraduates desiring to teach horticulture have no previous horticulture work experience. A similar problem exists for undergraduates with aspirations of teaching in small animal care, forestry, agricultural resources, agricultural business supply and service, agricultural mechanics, and, to some extent, production agriculture programs (7.25).

• • •

In agricultural education at The Ohio State University, a faculty member devotes the major portion of time and effort in working with students needing additional occupational experience. He/she aids in identifying their training needs and assists in placement in agricultural and natural resources businesses and industries as well as commercial farms in order to provide the practical experience needed. Training programs which are very flexible in nature, time of year, and duration are developed between the employer and student. The work is supervised and evaluated by the employer, personnel from the university, and cooperating teachers of vocational agriculture or county agents (7.26).

Internships or supervised occupational experiences offered through an agricultural education department should include (7.27):

1. An assessment procedure to identify the competencies needed and possessed by students.

2. Necessary guidelines, procedures, and forms to guide the conduct of the program.

3. Preparation of teachers of vocational agriculture as coordinators to assist in the supervision of students placed on internships.

4. Identified agricultural businesses and personnel capable and willing to provide internship experiences.

The operational principles by which the student internship operates at The Pennsylvania State University's College of Agriculture are (7.28):

1. That each student intern have at least one well defined work activity that is regarded as worthwhile by the cooperating organization with whom the intern is affiliated, the intern, and the faculty advisor.

2. That each intern develop specific learning objectives that can be readily identified and reviewed periodically throughout the work period.

3. That each intern be supported by a college related faculty person and cooperating agency representative. The roles of these support people are to assist with task identification, learning objective definitions, carrying out the task, counseling with the intern, and carrying through with ideas and projects initiated.

4. That each intern contract as an independent agent with the organization involved to do the work and pursue the learning objectives.

5. That regular meetings be scheduled to permit intern-to-advisor feedback and accountability.

6. That each intern assess the worth of the internship experience in a non-school setting and produce a report for the academic advisor illustrative of the learning realized through the experience.

Credit offered for supervised internship experiences ranges from none to one quarter-hour per week (40 to 50 hours) of internship. The agricultural education department at The Ohio State University provides one quarter-hour credit for each 100 hours of supervised internship experience.

Teacher educators should conduct internship programs in much the same manner as they expect teachers to supervise occupational experience programs. There should be a training plan, a training agreement, employer progress reports, and a record of activities in which the student participated.

Assessing Technical Competence

> Two basic strategies have been used for assessing the level of technical competence of beginning teachers of agriculture. Primarily these assessments are made at the time the prospective teacher applies for certification; however, in some cases a partial assessment is made at the time the prospective teacher formally enters a pre-service education program. For teachers who complete a pre-service teacher education program in a college or university, assessment of technical competence is largely in terms of the number of credit hours of course work completed in subject matter areas that comprise or are closely related to the instructional content that the teacher will teach once he or she is on the job. If prospective teachers have completed certain courses, they are assumed to possess at least the minimum level of technical competence needed to enter teaching (7.29).
>
> • • •
>
> A second approach to assessing technical competence is no more defensible than the "credit hour earned approach." "Years of work experience in an occupation" is used also as an indication of the level of technical competence (7.30).

Self-assessment

A quick and unscientific way of using existing competency lists to assess level of technical competence is self-assessment by students. Skills inventory checklists are used at the University of Minnesota and at The Ohio State University. Students rate themselves on specific tasks in order to identify areas for which occupational internship work experience or additional practical coursework is needed. At the University of Minnesota, students have the opportunity to compare their scores with an average of all students and an average of student teacher supervisors in each of the technical areas of livestock, soils, crop production, agricultural mechanics, farm management, agribusiness, horticulture, forestry, natural resources, and feed, grain, and fertilizer. Students rate themselves on a four-point scale as to whether they have performed the skill "well enough to instruct others," "without supervision," "with supervision," or "not at all." An advantage of the self-assessment procedure is that it informs students of the skills needed in teaching and encourages them to accept much of the responsibility to acquire the needed level of competency.

Application of Competency-based Strategy

The competency-based strategy can be applied to technical courses in agriculture. Teacher educators can work with the experts in the technical subject matter areas to identify more clearly the technical competencies that are to be developed in each course so that satisfactory completion of these courses can be interpreted more specifically as the acquisition of certain knowledges and skills needed by entering teachers (7.31).

Performance Tests

Levels of technical competence can be determined by the use of performance tests. Properly developed and validated, written tests can also be helpful in assessing a person's level of knowledge about technical subject matter. In addition to measuring knowledge, performance tests should be used to ascertain the prospective teacher's competency in mastering critical job skills. The development, validation, and administration of performance tests are difficult, time-consuming, and expensive tasks (7.32).

Documentation of Knowledge and Skills

> A fourth possible approach for assessing more precisely the technical competencies possessed by prospective teachers pertains specifically to the assessment of knowledge and skills acquired through experiential learning. A process whereby an evaluation can be made of learning from experience involves the prospective teacher preparing a "portfolio"--a petition describing the relevant knowledge and skills acquired from experience. The portfolio might include a description of the basic knowledge and skills acquired, an indication of how the teacher sees the knowledge and skills relating to the instructional content to be taught, and some documentation of the learning experiences. Documentation could include letters from employers, samples of work, and certificates indicating successful completion of on-job or industry training programs. Once the portfolio is prepared, then it is up to the appropriate persons to evaluate the petition to determine whether the prospective teacher possesses an acceptable level of technical competence (7.33).

Certification Based on Competence

Certification standards now specify college preparation in terms of courses and credit-hours, and occupational experience in terms of relatedness and months. More precise measurement is needed. Such measurement must be able to withstand legal challenges and provide a more reliable indication of competence in technical agriculture. Consideration needs to be given to what the scope, depth, and nature of occupational experiences should be and to whether "credit-hours earned" in coursework is a true indication of technical knowledge.

Implications

The previous discussion relating to the preparation of vocational agriculture teachers in essential knowledges and skills related to their areas of teaching suggests implications for programs of teacher education in agriculture. These implications are:

1. Technical preparation in agriculture should receive priority over general education and professional education in the allocation of credit-hours in the curriculum of the prospective teacher.

2. Agricultural education programs should maintain standards concerning the technical competence of entering vocational agriculture teachers.

3. Teacher educators should be able to interpret and apply basic principles as well as demonstrate essential skills.

4. Teacher educators should assume responsibility for assisting students in acquiring the needed level of technical competence through coursework and occupational experiences.

5. The curricula in colleges of agriculture should provide practical, hands-on experiences in agriculture for prospective teachers of vocational agriculture.

6. Teacher educators should encourage skill development and practical laboratory courses with reasonable class sizes in subject matter departments.

7. Prospective teachers should gain practical experience in agricultural occupations directly related to their intended teaching specialties.

8. Programs should be flexible to accommodate students who desire to prepare to teach at varying levels and in areas of specialization in agriculture.

9. Agricultural education departments, in conjunction with subject matter departments, should develop dual-major programs to prepare prospective teachers for specialized teaching roles.

10. Agricultural education departments, in approving university programs for teacher education in agriculture, should consider carefully the breadth and depth of agriculture programs that are available.

11. Teacher educators should support the need for modern livestock, greenhouse, agricultural mechanics, land laboratory, and experimental farm facilities for use by subject matter departments in preparing students for the real world of agriculture.

12. Teacher educators should develop and use lists of competencies in each specialty area to aid prospective teachers in planning their curricula.

13. Agricultural education programs should devise better measures of technical expertise to improve the concepts of "credit-hours earned" and "months of occupational experience."

14. Teacher educators should recognize that beginning teachers will not possess all the competencies needed for all the occupations for which vocational agriculture students are being prepared.

15. Teacher education in agriculture programs, in planning for vocational agriculture teacher needs, should give high priority to inservice education in technical agriculture areas.

CHAPTER 8

The Curriculum: Professional Education*

John R. Crunkilton
VIRGINIA POLYTECHNIC INSTITUTE AND STATE UNIVERSITY

Paul E. Hemp
UNIVERSITY OF ILLINOIS

The preparation of teachers of agriculture must include emphasis in many specific areas. These areas can generally be categorized into general education, agricultural subject matter and occupational experience, and professional education. Chapters 6 and 7 covered the first two of these areas, and this chapter is devoted to the professional education aspect of an undergraduate teacher education in agriculture program.

Professional education must be taken seriously by those who are associated with undergraduate teacher education programs. A student may have a sound general education background and may be expert in a specific taxonomy or subject matter area. However, neither of these conditions implies that a person possesses the pedagogical skills needed to direct the learning processes of students. The professional education portion of an undergraduate curriculum in teacher education must be a quality experience for prospective teachers. It must provide the pedagogical competencies needed by beginning teachers to conduct a local agricultural education program successfully.

The competencies and subject matter areas which should be a part of professional education in teacher education programs are explored in this chapter. The current state of these is discussed, and implications/issues for the future are presented. What will be included in the professional education aspect of the curriculum of individual institutions must rest with the respective staff therein.

---

*Material included in the first three sections of this chapter has been adapted from G. W. Weiger, Jr., "The Curriculum: Professional Education," which was Chapter VII in the 1967 edition of this book.

Historical Perspective

Contemporary teacher education in agriculture programs have evolved from combinations of general education, technical education (agricultural subject matter and occupational experience), and professional education. Professional education refers to that component of teacher education which is designed primarily to orient college students to the purposes, principles, policies, and procedures in education, as well as to develop in students the abilities essential in teaching vocational agriculture.

A major part of professional education is provided through formal college courses taught by professional teacher educators. These courses usually can be identified in an institution's list of offerings as "education," "vocational education," or "agricultural education" courses.

The professional education provided prospective teachers of agriculture is now universally accepted by those involved in preparing and supervising teachers of vocational agriculture, but professional education has not always been an integral part of the teacher education program.

Professional education (pedagogy) took root in American institutions of higher learning during the last two decades of the nineteenth century. By the turn of the twentieth century, proponents of professional education maintained that professional education courses should be included as a necessary part of the teacher preparation program. This movement relating to the training of teachers was a part of a larger educational reform movement (8.1).

At the same time, high schools in the United States were growing and changing in complexity. Administrators of these schools were beginning to regard areas such as social competence and vocational training as valid goals. With increasing enrollment in secondary schools, a wide range of student interests and abilities, and expanded curricula, teaching became a more demanding and complex profession. Many school administrators strongly expressed their views regarding the preparation of teachers for secondary schools. They agreed, along with the professional educators, that teachers needed more than content preparation and on-the-job experience to become competent teachers.

Professional educators and school administrators succeeded in translating their convictions relating to professional training into state certification regulations. Thereafter, institutions concerned with teacher education included in their curricula for teachers courses such as educational psychology, philosophy of education, and teaching methods.

Present-day teacher education in agriculture is built upon past achievements. A small milestone for teacher education in agriculture was the passage in 1907 of the Nelson Amendment to the act providing the annual appropriation for the U.S. Department of Agriculture. This amendment allowed for increases in federal funds appropriated for the support of colleges of agriculture. A proviso in the amendment made it permissible for colleges of agriculture to devote a part of the increased funds to "the special preparation of instructors for teaching the elements of agriculture . . . (8.2)."

Prior to 1907, practically the only effort in the training of teachers of agriculture was directed toward preparing elementary and, particularly, rural teachers to teach "nature study (8.3)." By 1912, 40 agriculture colleges offered a limited assortment of courses designed to prepare secondary teachers of agriculture. In some cases, curricula included only an elective or two in psychology and pedagogy (8.4).

At the end of the 1916-17 school year, only 19 colleges had programs of agricultural teacher education. The 1917 Smith-Hughes Act changed the picture quickly, for within two years after its passage, 21 other land-grant colleges had initiated such programs (8.5).

The Smith-Hughes Act gave the land-grant colleges a virtual monopoly on the professional education of teachers of vocational agriculture because federal funds provided could not be used in private institutions. This pattern remained until the passage of the Vocational Education Act of 1963.

In the early days of vocational agriculture, the curriculum for teachers of agriculture contained general education, technical education, and some professional education. The content and emphasis have changed through the years, but the overall framework has remained intact. It is still generally agreed that teachers of vocational agriculture need a broad, liberal, or general, education, a thorough understanding of and competence in agriculture, and high quality professional education. An assessment of views relating to the nature and content of professional education indicates less than full agreement between educators on what is desirable and what is essential.

When the Smith-Hughes Act was passed, there was a very limited supply of trained agriculture teachers. Practical farmers, secondary science teachers, and occasionally teachers of nature study were pressed into service. Weaknesses soon appeared in the qualifications of such teachers. They lacked the training needed to carry out the new vocational agriculture program. Graduates of agriculture colleges who had sufficient farm experience were available, but they too lacked the ability to produce the results desired from the teaching of vocational agriculture.

Nevertheless, the vocational agriculture program was launched with an assortment of teachers. As might be expected, after two or three years, some programs were discontinued because the teachers were not achieving desired results--not from lack of effort on the part of the teachers, but because they lacked ability in teaching methods (8.6). During the next few years, significant steps were taken toward providing teacher education on a more professional level.

The definition of professional education has been modified since the passage of the Smith-Hughes Act. Professional education was originally conceived as training given to occupationally competent persons which would render them capable of more effectively teaching agriculture on a vocational basis. An individual was considered occupationally competent if he/she was qualified to carry on production enterprises in agriculture successfully. The prospective teacher of vocational agriculture was expected to come into the program only after having had successful farming experience and having pursued or having begun college training in technical agriculture.

Early federal leaders in agricultural education had a concept of a successful farmer that has not been shared by some agricultural education leaders in recent years. The 1924 federal bulletin on teacher training in agriculture states:

> To be a successful farmer does not necessarily require ability in such things as organizing and teaching subject matter, writing for the rural press, public speaking, or proficiency in rural leadership. These are not essential to ability in effectively producing and marketing farm products--the main business of the farmer as such (8.7).

These early leaders felt that it was the purpose of the teacher education departments in the land-grant institutions to develop a good agriculture teacher out of a good farmer--not to make a good farmer out of a poor or partially equipped farmer--that was the job of the agriculture colleges.

Purposes of Professional Education

The primary purpose of professional education is to develop professional practitioners--teachers who perform professionally in fulfilling their responsibilities. Therefore, the professional education of teachers of vocational agriculture encompasses more than just hours of education credit required for certification. Each major part of the program of teacher education contributes to the development of a "professional" teacher; but professional education, perhaps, makes the most significant contribution.

A basis for developing professional education programs is a comprehensive understanding of the behavior of professional teachers. As characterized by Taylor (8.8), professional teachers of vocational agriculture:

1. Are alert to trends and need adjustments in the field. They maintain a positive attitude toward change.

2. Are interested in continuously improving their professional competence. They recognize the need for constant training and retraining. They read broadly, talk with leaders in various fields and are informed.

3. Take advantage of graduate education and the opportunity to work for an advanced degree. They participate in other inservice activities and read professional magazines and support their professional organizations.

4. Develop and maintain effective relationships. They become an effective part of the school and community.

5. Develop an understanding of the purposes and procedures of other vocational services.

6. Perform with a high degree of competence. They effectively plan and conduct a comprehensive program.

7. Are concerned with standards and quality, not only in their own departments, but also in the total program in the state and nation.

8. Focus on serving others. They are not concerned with self-aggradizement, perpetuation, or empire building but rather with improving the lives and circumstances of their students.

9. Recognize their professional obligations to identify and recruit capable young men and women for the agricultural education profession.

10. Maintain a positive attitude toward their job and their program. They believe in the future of agriculture and what they are doing. They tell others about it.

An analysis of the foregoing characteristics indicates that the professional phase of teacher education should consist of three distinct aspects: professional knowledge, professional attitudes, and professional skills.

Since the passage of the Vocational Education Act of 1963, more attention has been directed toward preparing teachers who can educate students for gainful employment in the broad field of agriculture. Not only must current preparation programs for teachers of vocational agriculture be different from those required for teachers of academic subjects, but they also must be different from earlier programs which were designed to prepare teachers to teach vocational education for farming. In 1965, Woodin identified competencies or qualifications needed by the new generation of vocational teachers, including vocational agriculture teachers, as being (8.9):

1. Understanding of career opportunities within the specific field and the ability to guide students in selecting appropriate career objectives.

2. Experience in, and knowledge of, the vocation and sufficient background to make an analysis of the tasks in which the worker must become proficient.

3. Ability to teach, on an individual and group basis, for occupational proficiency in the classroom.

4. Enough ingenuity to plan occupational experience programs which will prepare each student for initial employment or advancement.

5. Communication skills necessary for relating vocational education to the school and community, counseling with parents and placement in positions of employment.

Much of professional education can be done through on-campus professional courses and off-campus student teaching, but other professional experiences must also be provided. Comprehensive professional training programs include opportunities for prospective teachers to make personal contacts with teacher educators and other professionals outside the classroom. Additionally, students are provided the opportunity to participate in clubs and organizations designed to further their professional development.

Analyses of professional training programs indicate that the prospective agriculture teacher must learn to execute many tasks in order to be an effective teacher.

The most recent attempt to identify the professional competencies needed by agriculture teachers was conducted in 1977 when standards for agriculture/agribusiness were developed. Specific discussion on standards relating to teacher education are included later in this chapter.

Theory vs. Practice

Professional education has been criticized as being too theoretical. Some individuals think that professional education courses contain too much theory as opposed to practice or that theory is not applicable to the subsequent practice.

What is it about theory that causes critics to make negative statements? Theory in education has to do with the meanings, symbols, and general principles that underlie and explain relationships between educational phenomena. The theory of professional education stresses the ideas, generalizations, hypotheses, and assumptions underlying educational processes. It is important that prospective teachers should have knowledge about the fundamental ideas that explain the tasks successful teachers must perform.

Being able to perform a teaching task is not the same as being able to do the task with an understanding of why the task is being executed. Teachers lacking a theoretical base in education come closer to being technicians than professional teachers.

Prospective teachers of agriculture should learn early in their training that much of the theory of education will not apply directly to their daily tasks in teaching. Broudy states that "foundational knowledge . . . is used interpretatively as precise but large-scale cognitive maps on which programs are plotted but not solved (8.10)." He continues by stating that "for the solving of problems, i.e., for the applicative use of knowledge, theory has to be supplemented by technology, and only the specialist (who has both) uses knowledge applicatively."

Theory and practice can be distinguished in function, but they are closely related. Competent teacher educators plan for theory and practice to be experienced as close together as possible in the professional training program. Teacher educators have the responsibility of helping prospective teachers of vocational agriculture to apply theory to practice.

What professional education courses are theoretical in nature? Professional education courses of concern to prospective teachers of vocational agriculture include general professional education courses, agricultural education courses, and related courses. Some general professional education courses available to agricultural education majors which are considered highly theoretical are the historical, psychological, philosophical, and social foundations in education. Social science courses are related to education courses. These include history, psychology, economics, and sociology.

Most on-campus undergraduate courses in agricultural education are quasi-theoretical in nature. The content is organized in terms of distinctive problems of practice, and much emphasis is placed on "how to do" varied tasks in skillfully conducting vocational agriculture programs. A few courses in agricultural education have professionalized content where methods and technical subject matter content are integrated.

If professional education is liable to criticism as being too theoretical in training academic teachers, agricultural education may be open to criticism for not having enough educational theory. Most agricultural education majors have limited background in the history and philosophy of education. They also have little education in the social sciences and the "societal" foundations in education. This lack of background is reflected in the scores made by agricultural education majors on some National Teachers Examinations.

Standards for Teacher Education Programs

Any discussion of the professional education of prospective teachers of vocational agriculture should be related to the standards for quality programs in agricultural/agribusiness education (8.11) that were developed in 1977. One section of the standards directly relates to professional education in the curriculum. Of the many different sources and literature applicable to what is included in professional education, the recent standards and the manner in which they were developed reflect the best current thinking of the agricultural education profession regarding what is required in order to become a competent vocational agriculture teacher. Of the 97 standards which became a part of the standards for quality relating to teacher education programs, four standards related specifically to the curriculum in professional education.

These four broad standards do not include all topics associated with professional education. Other standards relate to special topics in other chapters in this book, specifically, the student teaching experience which is covered in Chapter 9.

In 1977, an AATEA ad hoc committee was set up to study standards as they related to teacher education programs. The results of this study are reported in Table 8.1 (8.12). Of the institutions responding, 92 percent supported Standard 12, which deals with the social, psychological, and educational foundations as an integral component of the instruction program. However, when these institutions rated their present programs, only about 58 percent of the respondents rated their programs as "good." When the percentage of non-respondents was subtracted, this left 39 percent rating their programs as "fair" or "poor" on this standard. Thus, more effort needs to be put forth to strengthen the social, psychological, and educational foundations in the instruction program.

Standard 13 concerns the field experiences and methods courses that are a part of the field experiences. Almost 99 percent of the institutions responding indicated support for that standard, and nearly 76 percent rated their programs as being "good." This indicates that most institutions feel they are meeting that standard. However, the other 24 percent must make a special effort to improve their programs to meet this accepted standard.

TABLE 8.1--Support for selected instructional program standards (Standards 12, 13, 14, 15, and 83) relating to professional education and a self-rating of the present teacher education programs

	Support for Standard				Rating of Present Programs						
Standard	Total N	Yes N %	No N %	Missing Cases N %	Total N	Good N %	Fair N %	Poor N %	No N %	N/A N %	Missing Cases N %
12. The social, psychological, and educational foundations are required as an integral component of the instructional program.	65	60 92.3	4 6.2	1 1.5	66	38 57.6	23 34.8	3 4.5	0 0.0	0 0.0	2 3.0
13. Field experiences and methods courses in agricultural education are closely integrated in the professional preparation of teachers of agriculture/agribusiness.	65	64 98.5	0 0.0	1 1.5	66	50 75.8	14 21.2	0 0.0	1 1.5	0 0.0	1 1.5
14. Students enrolled in the teacher education program develop the following teaching skills:											
a. Using the problem solving method.	65	64 98.5	1 1.5	0 0.0	66	45 68.2	19 28.8	2 3.0	0 0.0	0 0.0	0 0.0

(Continued)

TABLE 8.1 (Continued)

Standard	Support for Standard				Rating of Present Programs						
	Total N	Yes	No	Missing Cases	Total N	Good	Fair	Poor	No	N/A	Missing Cases
		N %	N %	N %		N %	N %	N %	N %	N %	N %
b. Planning and conducting demonstrations.	65	65 100.0	0 0.0	0 0.0	66	55 83.3	10 15.2	1 1.5	0 0.0	0 0.0	0 0.0
c. Planning and conducting field trips.	65	64 98.5	1 1.5	0 0.0	66	47 71.2	16 24.2	2 3.0	0 0.0	0 0.0	1 1.5
d. Using community resources.	65	65 100.0	0 0.0	0 0.0	66	43 65.2	21 31.8	2 3.0	0 0.0	0 0.0	0 0.0
e. Providing individualized instruction.	65	65 100.0	0 0.0	0 0.0	66	33 50.0	28 42.4	4 6.1	0 0.0	0 0.0	1 1.5
f. Using teaching aids effectively.	65	65 100.0	0 0.0	0 0.0	66	48 72.7	16 24.2	1 1.5	0 0.0	0 0.0	1 1.5

(Continued)

TABLE 8.1 (Continued)

Standard	Support for Standard				Rating of Present Programs						
	Total N	Yes	No	Missing Cases	Total N	Good	Fair	Poor	No	N/A	Missing Cases
		N %	N %	N %		N %	N %	N %	N %	N %	N %
g. Guiding students in selecting, planning, and conducting occupational experience programs.	65	65 100.0	0 0.0	0 0.0	66	38 57.6	23 34.8	4 6.1	0 0.0	0 0.0	1 1.5
h. Supervising occupational experience programs.	65	65 100.0	0 0.0	0 0.0	66	40 60.6	20 30.3	4 6.1	1 1.5	0 0.0	1 1.5
i. Organizing and using the FFA to strengthen the instructional program.	65	64 98.5	1 1.5	0 0.0	66	52 78.8	11 16.7	2 3.0	0 0.0	0 0.0	1 1.5
j. Applying learning principles in teaching.	65	65 100.0	0 0.0	0 0.0	66	46 69.7	18 27.3	1 1.5	0 0.0	0 0.0	1 1.5
k. Organizing and teaching postsecondary students and adults.	65	63 96.9	2 3.1	0 0.0	66	19 28.8	30 45.5	11 16.7	3 4.5	2 3.0	1 1.5

(Continued)

TABLE 8.1 (Continued)

Standard	Support for Standard				Rating of Present Programs						
	Total N	Yes	No	Missing Cases	Total N	Good	Fair	Poor	No	N/A	Missing Cases
		N %	N %	N %		N %	N %	N %	N %	N %	N %
l. Organizing and advising FFA, YFA, and other leadership developing organizations.	65	64 98.5	1 1.5	0 0.0	66	42 63.6	19 28.8	4 6.1	0 0.0	0 0.0	1 1.5
m. Organizing programs for educating the handicapped and other special need groups.	65	55 84.6	10 15.4	0 0.0	66	7 10.6	18 27.3	34 51.5	2 3.0	1 1.5	4 6.1
n. Providing career awareness and exploration programs in elementary and middle school grades.	66	54 83.1	10 15.4	1 1.5	66	10 15.2	20 30.3	23 34.8	2 3.0	5 7.6	6 9.1

(Continued)

TABLE 8.1 (Continued)

Standard	Support for Standard				Rating of Present Programs						
	Total N	Yes	No	Missing Cases	Total N	Good	Fair	Poor	No	N/A	Missing Cases
		N %	N %	N %		N %	N %	N %	N %	N %	N %
15. Students enrolled in the teacher education program acquire the following management skills:											
a. Organizing and using local advisory committee.	65	65 100.0	0 0.0	0 0.0	66	34 51.5	23 34.8	8 12.1	0 0.0	0 0.0	1 1.5
b. Determining community and student needs.	65	65 100.0	0 0.0	0 0.0	66	37 56.1	27 40.9	2 3.0	0 0.0	0 0.0	0 0.0
c. Developing annual and long-range program plans.	65	65 100.0	0 0.0	0 0.0	66	35 53.0	25 37.9	5 7.6	1 1.5	0 0.0	0 0.0
d. Developing a course of study for the local department.	65	64 98.5	1 1.5	0 0.0	66	51 77.3	13 19.7	1 1.5	1 1.5	0 0.0	0 0.0

(Continued)

TABLE 8.1 (Continued)

Standard	Support for Standard				Rating of Present Programs						
	Total N	Yes	No	Missing Cases	Total N	Good	Fair	Poor	No	N/A	Missing Cases
		N %	N %	N %		N %	N %	N %	N %	N %	N %
e. Arranging for adequate occupational experience programs.	65	65 100.0	0 0.0	0 0.0	66	36 54.5	26 39.4	4 6.1	0 0.0	0 0.0	0 0.0
f. Determining and acquiring needed instructional materials.	65	65 100.0	0 0.0	0 0.0	66	41 62.1	24 36.4	1 1.5	0 0.0	0 0.0	0 0.0
g. Determining and acquiring needed facilities and equipment.	65	64 98.5	1 1.5	0 0.0	66	30 45.5	31 47.0	3 4.5	2 3.0	0 0.0	0 0.0
h. Developing and using a filing system.	65	63 96.9	2 3.1	0 0.0	66	29 43.9	21 31.8	13 19.7	2 3.0	0 0.0	1 1.5
i. Organizing and operating a multiple-teacher department.	65	63 96.9	2 3.1	0 0.0	66	19 28.8	30 45.5	14 21.2	2 3.0	1 1.5	0 0.0

(Continued)

TABLE 8.1 (Continued)

Standard	Support for Standard				Rating of Present Programs						
	Total N	Yes	No	Missing Cases	Total N	Good	Fair	Poor	No	N/A	Missing Cases
		N %	N %	N %		N %	N %	N %	N %	N %	N %
j. Evaluating students and programs.	65	64 98.5	1 1.5	0 0.0	66	38 57.6	27 40.9	1 1.5	0 0.0	0 0.0	0 0.0
k. Completing and analyzing annual reports.	65	61 93.8	4 6.2	0 0.0	66	23 34.8	25 37.9	14 21.2	2 3.0	1 1.5	1 1.5
l. Articulating secondary and postsecondary programs.	65	57 87.7	8 12.3	0 0.0	66	8 12.1	30 45.5	20 30.3	3 4.5	2 3.0	3 4.5

(Continued)

TABLE 8.1 (Continued)

Standard	Support for Standard				Rating of Present Programs						
	Total N	Yes	No	Missing Cases	Total N	Good	Fair	Poor	No	N/A	Missing Cases
		N %	N %	N %		N %	N %	N %	N %	N %	N %
Instructional Program											
83. The agricultural education curriculum provides for adequate preparation for teachers of vocational agriculture/agribusiness with course work distributed as follows: General Education: 20-30% Professional Education: 20-30% Technical Agriculture/ Agribusiness: 20-40% Electives: 10%	65	56 86.2	5 7.7	4 6.2	66	12 18.2	3 4.5	32 48.5	13 19.7	6 9.1	

Source: An Analysis of Standards for Teacher Education Programs . . . (8.12).

Data on Standard 14 reflect how institutions rate the development of special skills needed by their prospective teachers of vocational agriculture. For example, teacher education programs do not meet Standard 14k, "organizing and teaching postsecondary students and adults," Standard 14m, "organizing programs for educating the handicapped and other special need groups," and Standard 14n, "providing career awareness and exploration programs in elementary and middle school grades." These data clearly indicate that teacher educators in agriculture need to revise the professional education aspect of teacher education programs to prepare teachers who can meet the current needs of society and the education programs that exist within our schools.

The data in Standard 15 indicate that developing management skills in prospective teachers should become a high priority for teacher education programs. It is evident by comparing the data associated with Standard 14 and Standard 15 that teacher educators feel they are doing a much better job of developing the basic teaching skills required of prospective teachers than they are of developing skills in the management area.

Another aspect of the professional education of prospective agriculture teachers deals with the overall curriculum in the undergraduate program.

Standard 83 addresses this as follows:

> The agricultural education curriculum provides for adequate preparation for teachers of vocational agriculture/agribusiness with course work distributed as follows:
>
> General Education 20-30%
> Professional Education 20-30%
> Technical Agriculture/Agribusiness . . 20-40%
> Electives 10%

About 86 percent of the institutions reported that Standard 83 was important for teacher education programs. However, only about 23 percent of the institutions responded that they either met or exceeded the standard, and nearly 49 percent said they did not meet the standard. While it is not possible to pinpoint which of the four categories of the standard institutions are not meeting, it can be assumed that the professional education core is an area that needs to be studied in more detail to find out if institutions do have 20 to 30 percent of their curricula set aside for professional education.

There is a consensus among teacher educators that the standards are important to teacher education programs. However, when the data are analyzed regarding the situation in current programs, there are some differences as to the extent to which the standards are currently being met.

These findings suggest the following implications for teacher education programs:

1. Additional research is needed to determine to what degree the standards are being met nationally, as well as in individual institutions. While some institutions may meet all the standards, there are others that would need to modify their programs in order to achieve the level of the standards.

2. Further study should be devoted to determine what percentage of instructional programs are allocated to general education, professional education, technical agriculture/agribusiness, and electives. While the standards generally reflect the current thinking of professional educators, some flexibility must be built into the programs.

3. Regardless of national statistics, each institution needs to apply the standards to its own program and then make an assessment as to areas of strength and areas of weakness that need to be modified in the future. Overall improvement of teacher education programs on a national level will only be accomplished by each institution's improving its respective program.

4. Once an institution determines that a program area needs to be strengthened to meet the standard(s), the next step is for it to identify goals and strategies that will accomplish upgrading. This has to be done on an individual basis, and the strategies identified would need to be based on the situation that exists within an institution at that time.

Pedagogy Currently Taught

In 1977, the AATEA appointed an ad hoc committee on teaching techniques to study the methods courses that were being offered to prospective teachers of vocational agriculture (8.13). The results of this committee's efforts were reported at the 1977 AVA convention in Atlantic City. Selected aspects of the data collected are presented here for discussion. It was reported that most institutions offer methods courses to their students prior to student teaching. It was also found that approximately 93 percent of the institutions offer methods courses only to agricultural education students. This is a strength of teacher education in agriculture programs in that instructors in these courses have the opportunity to help students apply the methods taught to specific areas of agriculture.

The study also identified which teaching techniques were being taught in the methods courses. Table 8.2 shows that demonstrations, questioning, discussion, field trips, supervised study, and resource persons are included in three out of every four institutions. However, other teaching techniques are taught less frequently. Educational TV, debates, experiments, games, team teaching, and programmed instruction are taught by less than half the institutions reporting. This is interesting since the agriculture teacher has many opportunities to use these teaching techniques. Almost every school has a mechanics laboratory, and many departments now have greenhouses and land laboratories. Thus, there is ample opportunity for agriculture teachers to use varied teaching methods.

Since methods courses are an important part of undergraduate programs, it is important to assess the attitudes of the profession as to what topics should be included within the methods courses. The ad hoc committee raised this question with teacher educators across the country and determined the degree of agreement as to which pedagogical areas should be taught. This information is summarized in Table 8.3. The statements falling in the strongly agree category (means of 1.54 and lower) reflect specific topics that

TABLE 8.2--Number and percentage of institutions including instruction in methods courses on specific teaching techniques

Teaching Techniques	Number (N = 78)	Percent
Demonstrations	78	100
Questioning	76	97
Discussion	76	97
Field trips	75	96
Supervised study	73	94
Resource persons	73	94
Lectures	67	86
Panel discussions	63	81
Student reports	62	79
Role playing and/or skits	57	73
Pre-tests	52	67
Brainstorming	52	67
Videotapes	51	65
Programmed instruction	44	56
Team teaching	41	53
Games	40	51
Experiments	38	49
Debates	28	36
Educational TV	24	31

Source: An Assessment of the Pedagogical Skills Taught to Agricultural Education Undergraduates (8.13).

TABLE 8.3--Areas of instruction in methods courses and degree of agreement as to whether the areas should be taught

Areas of Instruction	Area Should Be Taught in Methods Courses (number of responses)				Total Responses	Mean[b]
	SA[a]	A[a]	D[a]	SD[a]		
Lesson planning	74	3	0	0	77	1.03
Teaching techniques	75	4	0	0	79	1.05
Motivating students	72	7	0	0	79	1.08
Teaching strategies or approaches	68	11	0	0	79	1.13
Desirable teacher characteristics	63	12	4	0	79	1.25
Principles of learning	51	24	1	1	77	1.37
Developing behavioral objectives	50	28	1	0	79	1.37
Grading	47	30	1	0	78	1.41
Test construction	42	34	2	0	79	1.48
Teaching calendar or course of study development	50	16	12	1	79	1.54
Supervised occupational experience programs	45	14	11	3	73	1.61
Preparation of audiovisuals	39	32	7	1	79	1.62
Program planning	45	15	15	4	79	1.72
FFA	45	14	13	6	78	1.74
Curriculum materials development	35	29	12	2	78	1.75
Filing systems	26	37	10	3	76	1.86
Agricultural education facilities	26	31	15	6	78	2.01
Young farmer and adult education	27	26	13	8	74	2.02
Advisory councils	28	29	13	9	79	2.03
History and philosophy of agricultural education	14	37	16	10	77	2.28
4-H	1	17	34	27	79	3.10

[a]SA--strongly agree; A--agree; D--disagree; SD--strongly disagree.

[b]Means were calculated by assigning the following values: SA = 1, A = 2, D = 3, and SD = 4, summing the scores, and then dividing by the number of responses.

Source: An Assessment of the Pedagogical Skills Taught to Agricultural Education Undergraduates (8.13).

deal with methods of teaching. However, those topics that appear on the higher portion of the scale (means of 1.61 to 3.10) are topics that many times are found in separate courses other than methods, for example, program planning, FFA, audiovisuals, and young farmer and adult programs. It is evident from the data that teacher educators believe that a certain body of knowledge should be taught within methods courses.

If the highly rated topics are not covered within the current methods courses in a program, then the institution should explore adding new courses to the curriculum that would cover these topics. Other alternatives would be to provide these topics via seminars while the student is student teaching, or by courses that are held concurrently with the student teaching field experience. The topics could also be covered in first-year teacher courses or follow-up courses to student teaching. The point to be made is that regardless of the method or alternative used to deliver these topics, the areas of instruction with means of 1.03 to 1.54 identified in Table 8.3 are critical topics and should appear in the professional education curriculum. The only topic in the table that could be questioned is 4-H. However, in some institutions across the country, agricultural education and extension are under the same administrative unit. Thus, students enrolled in such programs may be focusing upon extension education, and 4-H then becomes a very relevant topic.

Competency/Performance-based Programs

Many teacher educators in agriculture use the terms "competency-based teacher education" (CBTE) and "performance-based teacher education" (PBTE) interchangeably. Students enrolled in a competency-based teacher education program are expected to demonstrate essential teaching tasks in a real teaching situation. Students are also expected to have the necessary knowledge needed to carry out particular teaching tasks and to explain why they performed the tasks in a particular fashion.

The essential characteristics of the competency-based program include the following (8.14):

1. Specified competencies which students are expected to demonstrate are derived from a systematic analysis of performances of recognized practitioners. These competencies are stated in advance of instruction in measurable terms and made known to the student.

2. Evidence of the learner's achievements is obtained through assessment of learner performance, using criteria which have been stated in advance of instruction.

3. The instructional program is focused on the development of specified competencies and results of performance tests are used to guide individual learner's efforts and to determine his/her rate of progress and completion of the program.

These characteristics suggest that CBTE programs involve individualized, personalized, and modularized instruction; performance feedback; a systems approach; and exit requirements based on the demonstration of specified competencies.

Competency-based teacher education is seen by its advocates as a process for improving the preparation and development of educational personnel, including teachers, supervisors, counselors, and administrators. Broudy, writing in the November 1975 issue of The Journal of the American Association of Teacher Educators in Agriculture, summarizes the advantages of CBTE as follows (8.15):

1. The attempts to define and behavioralize goals will help teacher education institutions to clarify and justify their priorities and eliminate overlapping components of their programs.

2. Attempts to define essential teaching competencies should lead to greater analysis and possibly a better understanding of the teaching process; this may lead to the discovery of new instructional methods.

3. Specification of goals in terms of behavior should help the student and the public see what the educational program hopes to achieve and to evaluate these efforts.

4. More attention would be directed to the relation of theory to practice by the emphasis on "clinical" experiences.

Hillison summarizes the advantages of CBTE programs as follows (8.16):

1. Students are permitted to work at their own pace.
2. Emphasis is on exit, not entry, skills.
3. Objectives are made public.
4. Grades are not given.

Looking at the other side of the issue, Hillison also lists three disadvantages of CBTE programs. The disadvantages most often stated are that CBTE programs do not have a research base, that they are often inhumane and mechanistic, and that they take control away from the teacher educator.

Broudy raises questions about the efficacy of competency-based teacher education as a prime model for teacher education. He questions the assumption that all teaching and learning can be organized and evaluated on a model which calls for breaking down a complex operation into component parts and sequences and then practicing these components to achieve a level of mastery learning (8.17). This approach, while useful in training skilled workers and technicians, may not be appropriate or useful in preparing professional educators. The teaching-learning act cannot be broken down into observable performances without an interpersonal dynamics remainder. If competency-based teacher education deals only with the observable performances, the "remainder" will be ignored. Successful teachers teach using knowledge that affects the learning process. In agriculture, teachers draw upon a vast amount of knowledge about their students' home situations, their students' vocational goals, and their own previous experiences in agriculture. This knowledge is important to success in the classroom; however, it is not a part of the observable behaviors which are emphasized in the CBTE model.

In spite of the controversy concerning the merits of competency-based teacher education, many states and teacher education institutions have adopted it as a model, or they have implemented certain elements of the model, in their teacher education programs. Peterson directed an AATEA-sponsored study in 1975 to determine to what extent teacher educators in agriculture had implemented the PBTE concept in their undergraduate programs (8.18). Responses from 43 teacher educator institutions indicated that 81 percent of these institutions had implemented the PBTE approach partially or completely in one or more undergraduate agricultural education courses. The areas of professional education and the number of institutions using the PBTE concept in each of these pedagogical areas are summarized in Table 8.4.

TABLE 8.4--Undergraduate courses or subject matter areas in which the PBTE concept is at least partially utilized

Course	Number
Philosophy	5
Teaching methods	28
Teaching methods in agricultural mechanics	11
Adult education	10
Program planning and curriculum	18
FFA	17
Supervised occupational experience programs	12
Student teaching	21
Guidance	5
Coordination techniques	2

Source: R. L. Peterson, PBTE in Agricultural Education . . . (8.18).

Institutions which reported complete adoption of the PBTE concept in their agricultural education programs in 1975 included The University of Arizona; the School of Agricultural Science, Fresno, California; the

University of Florida; Southern University of Baton Rouge, Louisiana; and The University of Nebraska. Institutions which reported that they had adopted the PBTE concept completely in one or more courses were Michigan State University; the University of Minnesota; the University of Missouri; South Dakota State University; Utah State University, and the University of Wyoming.

Other institutions reported to the AATEA Committee on Teacher Education in Agriculture Guidelines that they were implementing selected aspects of performance-based teacher education or doing the research necessary to identify teacher competencies.

What is the future of CBTE programs in agriculture? It is proposed that the widespread use of the CBTE/PBTE concept will depend on answers to the following questions.

- What is good teaching in vocational agriculture?
- What are the competencies necessary for successful teaching?
- How will teacher performance be evaluated, and who will do the evaluating?
- Should the CBTE concept be applied in total or only to certain courses or selected pedagogical areas?
- Can the CBTE approach prepare teachers who are competent in heuristic teaching, which emphasizes discovery methods, and philetic teaching, which promotes satisfactory human relationships?
- To what extent will competency-based teacher education reduce the role of a teacher from a professional to a technician?

The CBTE movement has been strongly supported by state teacher certification authorities, the American Association of Colleges for Teacher Education, and a number of teacher education institutions. It comes at a time when the public call for accountability reflects disenchantment with public education, which includes teacher preparation programs. This may explain the rapid adoption of the CBTE approach. However, given the preceding unanswered questions, agricultural educators should not view the CBTE movement as a panacea to cure all educational ills.

Critical Issues Relating to Professional Education

Despite the advances made in the development of teacher education programs and the delivery of appropriate instruction for prospective teachers of vocational agriculture, many issues remain as problem areas for further research and discussion. Some of these issues have plagued teacher educators for many year, while others are relatively new issues which have emerged from changes in educational systems and society as a whole. Some of the important issues facing the profession today are as follows:

1. <u>How much emphasis should be placed on theory and practice in the professional education component of a teacher education program?</u>

Learning by doing is recognized as a fundamental concept to be applied not only in the student teaching center but also in the on-campus professional education courses. Competency-based teacher education strategies are forcing teacher educators to include more laboratory work and more student practice in their professional education courses.

2. <u>When should professional education courses be scheduled in the undergraduate program?</u>

In some institutions, agricultural education students are required to begin their professional education component at the freshman or sophomore level; thus, professional courses are spread out over a four-year period rather than concentrated in the senior year.

3. <u>To what degree should teacher education programs be competency-based?</u>

Competency-based teacher education programs have been adopted in some states and in some universities, but a majority of institutions have not been inclined to use the CBTE/PBTE approach totally. It is unlikely that the CBTE concept will dominate the educational scene unless, or until, the disadvantages and shortcomings of this approach can be overcome.

4. <u>How should the responsibility for providing instruction and practice in teaching be divided between university staff members and teachers in the public schools?</u>

Public school teachers, acting through strong professional organizations, are demanding and receiving a stronger voice in how teachers are prepared and certified. Teacher educators and public school teachers and administrators will cooperate more closely in planning and conducting teacher education programs in the future.

5. <u>Who should direct teacher education in agriculture?</u>

In most states, teacher educators in agriculture, state supervisors and consultants, and secondary teachers of agriculture have worked together to conduct teacher education activities at the preservice and inservice levels. In states where teacher education in agriculture is conducted at more than one institution, there is a special need to coordinate the activities of all institutions and groups.

6. <u>Should the professional component of an agricultural education program be taught in specialized courses or as general vocational education courses?</u>

Some educators believe that agricultural education students can be taught effectively in general vocational courses rather than in specialized agricultural education courses. Opponents of this approach argue that vocational agriculture teachers need skills and knowledges which differ significantly from those used in other vocational areas.

7. <u>How should the professional education program be adjusted for students who lack a vocational agriculture and FFA background?</u>

   Many students currently enrolled in teacher education in agriculture programs have little or no background in FFA and/or supervised occupational experience programs in agriculture. Teacher educators have been forced to spend increasing amounts of instructional time on material which did not previously have to be taught when students had a strong background in vocational agriculture.

8. <u>To what extent should state supervisory staff and department of education staff be involved in teacher education programs?</u>

   Prior to the 1960's, state supervisors of vocational agriculture and program specialists at the national level were able to exert considerable influence on teacher education in agriculture programs offered in universities and colleges. In recent years, state staff members in most states have had limited influence over teacher education. Changes in reimbursement policies and changes in authority given to state supervisors and federal program specialists in agricultural education have decreased the decision-making role of these professionals.

9. <u>How can the professional education program be kept up-to-date and made more effective?</u>

   Should teacher educators rely on the experiences of cooperating teachers and others regarding what should be taught in the professional sequence? Should research be emphasized as the primary source of direction in curriculum modification? Answers to these key questions are needed.

The foregoing nine issues also need to be considered in the context of issues and concerns applicable to all teacher education. Some of these issues and concerns are as follows:

1. Programs lack individualization and personalization.

2. Programs include too little or too much theory and a consequent lack of applicability in the real world.

3. Programs place too little or too much emphasis on the liberal arts.

4. Programs include too many or too few school-based experiences.

5. Programs place too much emphasis on the present and not enough on the future.

6. Programs attempt to prepare candidates for a generic teacher role, rather than for a variety of roles.

7. Program design is based on the assumption that one can learn through an apprenticeship approach.

These issues and concerns merit the attention of teacher educators, cooperating teachers, and state staff members if teacher education in agriculture is to function effectively in the years ahead. These issues may or may not apply to all programs of teacher education in agriculture. Issues need to be identified and addressed on a state or an institutional basis, and priority should be given to those problem areas judged to be most acute. Furthermore, national teacher education in agriculture issues should be addressed by national professional organizations, such as the American Association of Teacher Educators in Agriculture.

CHAPTER 9

The Curriculum: Field-centered Experiences

R. Paul Marvin
UNIVERSITY OF MINNESOTA

Some type of field experience has been a part of teacher education programs since the earliest requirements of any pedagogical preparation for teachers. Vocational agriculture instructors have had a field experience component in their teacher education programs in every state in the United States since the establishment of teacher education programs. A review of the literature indicates that field-centered experience programs differ considerably in length, type, time, and methods of implementation. However, some basic purposes transcend all such field-centered experiences. Thus, the most common methods of achieving these purposes are described in this chapter.

Because the programs that vocational agriculture instructors are expected to conduct include supervised occupational experience programs as an essential teaching tool, it is not surprising that teacher education programs for vocational agriculture instructors include a similar field-centered experience requirement. Field-centered experiences which provide agricultural subject matter competencies are included in Chapter 7 and therefore will not be addressed here.

All field-centered experiences, including observation and other laboratory experiences, culminate in a period of student teaching organized to provide competency in the wide range of professional activities in which teachers should engage.

The student teaching phase of the teacher preparation program is almost universally accepted as the most important part of the professional education of teachers.

> Academic professors and professors of education are in complete agreement only on one point: that practice teaching, if well conducted is important. Aside from practice teaching and the accompanying methods course, there is little agreement among professors of education on the nature of the corpus of knowledge they are expected to transmit to the future teacher (9.1).

Teacher educators have considered various ways to enhance the student teaching experience. Two examples are (1) pre-student teaching experiences, including performance-based teaching methods, courses, and (2) the concept of clinical supervision, introduced by Cogan (9.2) and his associate at Harvard.

Organization of Field-centered Experiences

There are many patterns for organizing the field experiences involving pre-student teaching, various lengths of time for student teaching, and different supervision methods. Very little research has been done to determine if there is any one best pattern. In fact, there is no evidence to indicate the most desirable length for the student teaching experience. However, it is an area that has received considerable attention and exploration by educators. The report of study groups at the DeKalb Conference on Teacher Education suggested some principles that should provide a guide for selecting and organizing laboratory experiences (9.3).

A. An important principle is to have the laboratory experience contribute to the establishment of understandings and feed into a sequence so there is continuity of teacher growth.

B. It is highly desirable that experiences be presented in true life situations, and that they be accompanied with such guidance that the components of a good teacher are built in the stages of development. Caution is urged in use of artificial, "canned" experiences that cannot involve the total situation.

C. The laboratory experiences should lead from those more highly directed to those more complex ones which will be pursued with increased student initiative and use of the supervisor as a consultant.

D. There is need of exploration and experimentation in ways to provide profitable experience for the student teachers.

The national standards for quality programs in agricultural/agribusiness education (9.4) established a minimum of 10 weeks of student teaching in the area for which certification is to be granted, but when a committee (Peterson, et al.) surveyed 53 agricultural teacher education departments, only 40 percent, or 21 departments, were in agreement. Twenty-two institutions required fewer than 10 weeks, whereas 10 institutions required more than 10 weeks. The range was from 6 weeks to 18 weeks.

Pre-student Teaching Experiences

The major and immediate purpose for providing students with pre-student teaching experiences is to prepare them more effectively for student teaching. Another purpose is to make the students' study of professional education and psychology courses more realistic. Giving students the opportunity to observe and evaluate classroom student behavior helps to make real the concepts and principles presented in class readings and discussions.

Observation experiences and/or other experiences are provided throughout the undergraduate program prior to student teaching and are supervised by the agricultural education staff. Pre-student teaching experiences, sometimes referred to as "early experiences," in addition to those integrated with professional education courses are often provided during the summer.

In other instances, observation experiences are for parts of several days or for one or two days at a time over a period of several weeks. In 1966, suggested outcomes from pre-student teaching were (9.5):

A. To become familiar with the community and the agricultural situation in the community.

B. To visit high school students, young and adult farmers, and to become familiar with their programs and problems, and to gather information to use in planning units to teach during the student teaching period.

C. To become familiar with the school and the vocational agriculture department's organization, facilities, programs of work, and teaching calendars.

D. To determine, insofar as possible, the units to teach during the teaching period.

Since the passage of the Vocational Education Act of 1963, vocational agriculture has expanded to serve more areas than production agriculture. Consequently, Outcome B should be broadened to include agricultural supplying and processing/distributing.

Most teacher education departments conduct some form of seminar for orientation to student teaching. The orientation includes topics such as (1) insurance (both automobile and personal liability), (2) relationship of student teacher to school administration, (3) relationship of student teacher to school and community, (4) expectations of the teacher-training institution for satisfactory performance.

Pre-student teaching experiences may be provided the student in many forms and through various administrative arrangements. However the experiences may be provided, they should effectively contribute the overall pedagogical competencies needed by beginning teachers.

Wehrer (9.6) describes a program at Gannon College in Erie, Pennsylvania, where the secondary-level pre-student teachers are provided systematized field experiences, which serves the purpose of both screening and sensitizing students to the realities of a career in secondary education. Students can explore their interest in teaching by working with secondary students in local schools and other community agencies before beginning their student teaching. This program requires that each student demonstrate a minimal level of competency by working outside the college in community learning sites. Working with teenagers at these sites enables the students to indicate that they have not only the ability to get along with teenagers but also the capacity to adapt subject matter to the level of the learners. The systematized program allows students to follow several paths through different kinds of experiences. At each point, a student may decide to consider another career; at no point is a student locked into a career in education.

Admission to Student Teaching

Students usually apply for student teaching during their fourth or last year in the college program. The selection of students for teacher education is a critical part of the undergraduate teacher education program. Teaching is not for everyone, so teacher educators have the responsibility of screening students in and out of teacher education. In so doing, the welfare of the secondary students who are to be taught, as well as the welfare of the prospective teachers, should be considered (9.7).

Screening may take place at several points in a prospective teacher's preparation program, but the most common point is at the time the student applies for student teaching.

O'Kelly learned from a study of 72 teacher education institutions preparing teachers of vocational agriculture that (9.8):

> Most institutions place major emphasis on the satisfactory completion of a prescribed course of study as a basis for admitting students to student teaching. In the order of frequency of mention the following factors were listed as being considered important by the majority of the institutions reporting: (1) satisfactory completion of a prescribed course of study; (2) scholastic average on course work to date; (3) farming background; (4) professional attitude and observed behavior; (5) general qualifications as measured by subjective evaluation by teacher trainers; (6) moral character; and (7) grades received on previous agricultural education courses. Very little attention appeared to have been given to factors such as: (1) age; (2) guidance test scores; and (3) physical fitness report.

The current emphasis placed on equal rights and opportunities has limited the ability of institutions to be involved in screening prospective students who have physical disabilities. In general, student teachers must meet the health standards of teachers in the field, which does not limit prospective teachers except for conditions that might jeopardize the health and welfare of students they would be responsible to teach.

An increasing number of teacher education programs utilize a review panel and an interview to make recommendations concerning potential teaching success (9.9).

The University of Minnesota utilizes a "case conference" for making decisions in special cases concerning admission to student teaching (9.10).

Student Teaching Experiences

The major objectives of student teaching, as stated in the handbook from the University of Minnesota, are (9.11):

1. To demonstrate the ability to teach students in high school (grade levels 9-12) and adult or young farmer programs of instruction.

2. To observe a comprehensive secondary school program.

3. To describe the purpose of vocational agriculture in a comprehensive school curriculum.

4. To demonstrate the use of community activities and resources in a vocational agriculture program.

5. To demonstrate the ability to organize and operate a vocational agriculture department.

6. To demonstrate the ability to work effectively with all phases of a complete vocational agriculture program, i.e. classroom and laboratory instruction, S.O.E.P., FFA and adult and young farmer instruction.

7. To demonstrate the ability to function effectively in various non-teaching activities related to the operation of a comprehensive school program.

The development of the ability to conduct inschool classes and to provide adult instruction in agriculture/agribusiness should be the central focus and should occupy the majority of time during student teaching. The college supervisor and the supervisory teacher must agree on a mode of operation that will permit the student teacher to move from observation to a partial teaching load and finally to the experience of a full teaching load comparable to what the student teacher will experience as a full-time instructor.

The student teacher writes lesson plans and develops learning experiences for the classes to be taught. These experiences include planning field trips, keeping records, promoting the FFA, supervising occupational experience programs, evaluating students (grading), providing guidance, and working with parents.

The guides or handbooks for student teaching from Montana, Minnesota, North Dakota, and Ohio all provide checklists of activities for student teachers, supervising teachers, and college supervisors to use as a guide to possible activities that might contribute to the success of future teachers. There needs to be a definite plan and understanding of how these are to be used to assure that the experience is practical for a specific individual or situation.

Selecting Supervising Teachers

Supervising teachers have such a major influence in shaping the teaching style of future teachers that teacher education departments must regard the responsibility of selecting supervising teachers as one of the most important steps in the teacher preparation process.

Supervising student teachers is an added responsibility to a teacher's already heavy schedule. Although most institutions pay the supervising teachers a token honorarium, this payment should not be construed as adequate remuneration for the added time and energy required to serve as supervising teachers. Supervising teachers work with student teachers because they consider it a worthwhile professional responsibility. Some of the key factors in selecting a supervising teacher are as follows:

1. The teacher will accept the responsibility for giving assistance to student teachers, as indicated by a willingness to schedule time to assist the student teacher in preparing for classes to be taught and other activities to be carried out in evaluating his/her performance as a student teacher. The teacher is willing to assume responsibility for the student teacher's performance in a variety of teaching situations. Consequently, the supervising teacher agrees to observe the student teacher, critique daily, and offer encouragement.

2. The teacher has had successful teaching experience in a program of vocational agriculture/agribusiness, which is effectively contributing to the preparation of youth and adults for employment in agriculture as evidenced by:

    a. An emphasis on the vocational objectives of agriculture, so that the instruction is employment oriented.

    b. The curriculum's being a blend of (1) a supervised occupational experience program for each student, (2) experience-oriented classes, and (3) a comprehensive FFA program of activities.

    c. The scope and quality of courses offered in the department.

    d. The scope and quality of the supervised occupational experience programs of the class members.

    e. The regularity and effectiveness of on-site instruction (SOE program visits) provided students.

    f. The program's reflecting a continuing education dimension, providing instruction in agriculture/agribusiness to out-of-school youth and adults.

3. The teacher has demonstrated an interest in professional improvement, as evidenced by a continued record of participation in professional activities which include pursuit of an advanced degree, summer school attendance, workshop attendance, and attendance at professional meetings for teachers of vocational agriculture. The teacher also participates in the professional activities of the vocational agriculture teachers association.

4. The teacher has three years of teaching experience, or in certain cases, is in his/her third year of experience.

5. The teacher has at least one year of experience in the school system in which the student teacher will be placed.

6. The teacher is a fully certified instructor-coordinator of vocational agriculture.

7. The teacher demonstrates enthusiasm for the profession by encouraging young people to become teachers of vocational agriculture.

Selecting Student Teaching Centers

The national standards (9.12) have 97 percent agreement that cooperating schools at the secondary and post-secondary levels to be used for student teaching and other field experience should be selected by teacher educators in cooperation with the staff of the state agriculture/agribusiness education unit.

General criteria for selecting schools to serve as teaching centers include:

1. The administration of the school has approved the use of the vocational agriculture department of the school as a student teaching center and has authorized the instructor to serve as a supervising teacher.

2. There is demonstrated evidence that the school is accomplishing outstanding outcomes in the education of its clientele.

3. The school is practicing those methods and techniques of education that student teachers should have an opportunity to observe and acquire.

4. The school provides a favorable educational environment for teaching and for learning.

Specifically, the vocational agriculture department should have the following characteristics:

1. The physical facilities of classrooms, mechanics laboratories; greenhouses, etc., provide a satisfactory setting for carrying out an effective and functional program.

2. Equipment for carrying out the program is satisfactory, maintained in usable condition, and readily accessible for instructional use.

3. The content and organization of the instructional materials in the department (books, bulletins, charts, transparencies, films, filmstrips, slides, and specimens) are adequate for a department of vocational agriculture.

A further consideration in selecting student teaching centers is the attitude and availability of other support school personnel with whom teachers must learn to work. A successful teacher must learn to relate to the school administrators, counselors, secretarial staff, and custodians, in addition to other classroom teachers.

The student teaching experience should provide an opportunity for the prospective teacher to become involved in the school and the community. Contacts of seven institutions surveyed all required that the student teacher reside in the community for part of the time during the student teaching assignment.

The literature search revealed no instance where some form of contractual arrangements did not exist between the college and the local school serving as the teaching center. In some instances, the state departments of education are included in the contract arrangements. If the state departments were not included in the contracts, in all the cases reviewed, they were consulted and informed of centers to be used for student teaching.

It is a frequent practice for the college to reimburse the local school system for its participation in student teaching experiences. The reimbursement is usually in terms of a fixed fee from as little as 50 dollars for each student teacher up to 200 dollars or more, depending on length and involvement of experience programs. The money may be paid directly to the school or to the supervising teacher. Whatever reimbursement procedure is followed, it must have the approval of the local school board.

A study of the literature reveals that most educators are of the opinion that supervising teachers should receive a substantial increase in salary for their supervisory services.

The Role of the Supervising Teacher

The role of the supervising teacher is one of great challenge and responsibility. Student teachers will turn to their supervising teachers for ideas, guidance, criticism, and commendation for years to come. Supervising teachers influence the professional attitudes and habits of the student teachers, as well as provide the opportunity for them to develop professional competencies.

Luft compiled the following composite of the supervising teachers' role from the handbooks of several states (9.13). Supervising teachers:

1. Respond immediately to the student teachers' initial contact, answering any questions they may have and outlining the classes and units they most likely will teach. Students should be given a general idea of the amount of time they will be expected to teach for each unit. This will necessitate some forward planning so that students can begin to make preliminary plans.

2. Lend assistance to the student teacher in obtaining suitable housing if such assistance is requested.

3. Prepare the classes for an incoming student teacher by briefly explaining what the student teaching program is about, and that the student teacher will be considered as part of the faculty and should be treated by students as such.

4. Introduce the student teacher to the administrator(s) immediately, and to other faculty members and school employees.

5. At the first opportunity, the supervising teacher and the student teacher should review school policy particularly on such matters as:

   a. Daily work hours.
   b. Weekends in the community.
   c. Responsibilities.
   d. Dress.
   e. Mileage and other expenses.
   f. Personal conduct.

6. Acquaint the student teacher with the school and Vocational Agriculture Department plant and facilities, and with the activities and procedures of the school and department. This undoubtedly will be a progressive undertaking, but student teachers should be familiar with such items if they are to benefit most from their experiences in the student teaching center, and if they are to be productive members of the teaching staff in Vocational Agriculture.

7. Plan ahead, set up, and discuss with the student teacher a schedule of teaching assignments and responsibilities for FFA and other activities for the entire six-week period. This will allow the student teacher sufficient time to plan ahead and schedule activities.

8. Plan student teaching activities so that the student teacher will gradually accept responsibilities. For at least one full week of his/her experience, the student teacher should assume a full load teaching to gain an exposure to the total job of teaching Vocational Agriculture.

9. Cooperatively plan the student teacher's daily activities and responsibilities early enough so he/she will have adequate time for preparation.

10. Coordinate lesson plan assignments with departmental course outlines, but if possible, start the student teacher with a unit in an instructional area in which he/she is well qualified.

11. Check lesson plans with the student teachers before they teach any class, but let them handle the class when they do their teaching--give them an opportunity to develop confidence in themselves.

12. Make it a point to take the student teacher on visits to students' occupational experience programs early in the student teaching period. The student teacher should have an opportunity

to observe the supervising teacher conduct a supervised occupational experience program visit.

13. Review the student teacher's assignments to become familiar with them so that advice, counsel, and encouragement may be given.

14. After the student teacher has become established with the classes being taught, make careful evaluations and offer constructive criticism. This should be done as soon after the teaching as possible. Provide time each day for conferences with the student teacher.

15. Observe the ability of each student teacher to work with other people in the school and community. The development of the ability to follow professional procedures and to get along well with people is as important for the student teacher as is the development of skill in classroom teaching.

16. Provide an opportunity for the student teacher to become involved in activities of the total school system and community, as well as those of the Vocational Agriculture program.

17. Submit a final evaluation report on each student teacher as required by the teacher education institution.

The Role of the College Supervisor

It is essential that the efforts of the college supervisors and the vocational agriculture instructors who serve as supervising teachers in student teaching centers be closely coordinated (9.14).

This communication of expectations is accomplished in several ways. A few institutions require the supervising teachers to have a credit course dealing with supervision of student teachers. Other institutions use the summer conference or special workshops to assure that there is agreement on the expectations for all persons concerned.

The role of the university teacher educator is to help the student teacher have a professionally rewarding student teaching experience. The teacher education institution places a high priority on student teaching. Therefore, professors who have taught the undergraduate courses and have guided students in their programs usually serve as the supervisors of student teaching. One of the key functions of the university teacher educator and the supervising teacher is to relate the basic principles of teaching vocational agriculture to actual practice. The college teacher educator acquaints the supervising teacher with the responsibilities assumed in working with a student teacher. The college teacher educator assumes responsibility for assigning a grade to the student teaching experience but will seek the advice and recommendations of the supervising teacher.

Evaluating the Student Teaching Experience

The Phi Delta Kappa Commission on Evaluation considers evaluation in light of the decision-making process. "Evaluation is the process of delineating, collecting and providing information useful for judging decision alternatives (9.15)."

The decision alternatives in student teaching are concerned with actions that will assist future teachers to improve their teaching competence and provide a basis for making employment recommendations to schools seeking instructors.

In 1967, Beamer listed four points as guides for evaluation of student teaching that are still valid today (9.16). These are:

1. Evaluation should be a cooperative effort by the student teacher and all who have supervisory functions. It should be performed on a continuous basis. The method for these evaluations should be objective evidence on criteria established cooperatively by the student teacher and the supervisory personnel.

2. Evaluation should be made not only of classroom teaching, but also of all other participating experiences included in the student teacher's program of student teaching.

3. Marking is the responsibility of the college granting the credit. However, evaluation by the supervising teacher should be given major consideration in arriving at a mark.

4. Every effort should be made in the evaluative process to bring the student teacher to the point where he has the competence and the desire to evaluate his own efforts.

The teacher education institutions provide inservice education for beginning teachers in their first year of teaching. Several institutions have established courses specifically for first-year teachers. In a few states, the course for beginning teachers is a graduate credit course meeting 6 to 10 times during the academic year. The course is designed to deal with the problems faced by the beginning teachers in their teaching. A second benefit is the opportunity for the teachers to begin a graduate program with this first course toward professional advancement.

The beginning teacher course also provides excellent evaluative input into the total teacher preparation program, including the student teaching experience.

Other Field Experiences

Internships

Several institutions have established internships for future vocational agriculture instructors. The purpose is to provide experiences for the prospective teacher that will enhance the student teaching experience and that may not be available during student teaching.

Purdue has a summer intern program which is designed to provide students with experience in planning, developing, and operating a vocational agriculture program during the summer months (9.17). Direct supervision is by a local vocational agriculture instructor, with overall direction provided by the teacher education staff in cooperation with the state agricultural/agribusiness unit. A specialized program for each apprentice is developed. Summer activities include making supervised occupational experience visits, attending FFA camps and conventions, working with out-of-school youth and adults in agriculture and attending professional workshops and conferences. The intent of this experience is to increase the number of agricultural education students who enter the teaching profession and to provide participating students with practical on-the-job experience in conducting vocational agriculture programs.

In addition to the intern program for teaching experiences, an increasing number of states have included a work experience internship as a requirement to be a coordinator of student occupational experience programs. Most institutions include coordinating techniques as part of the required courses. The purpose of the work internship is to provide the future coordinators with experiences similar to those of the students they will supervise. The work experience internship is supervised by the college supervisor.

Hemp suggests a number of other out-of-class activities that contribute to the preparation of future teachers of agriculture (9.18).

> Students should be expected to participate in a variety of out-of-class activities to better prepare themselves as teachers. Some of these activities are as follows:
>
> - Participate in the National Student Teacher Conference and the Alpha Tau Alpha Conclave.
> - Participate in student clubs, such as Alpha Tau Alpha and Agricultural Education Clubs.
> - Teach an agriculture unit to a class in elementary school (Food for America Program).
> - Serve as assistant advisors in various sectional and state FFA contests and award programs.
> - Observe and study vocational programs in agriculture for special needs students, community college students, urban students, etc.

CHAPTER **10**

Student Personnel Services in Teacher Education

Charles W. Byers
UNIVERSITY OF KENTUCKY

A comprehensive review of the goals of higher education since the first days of the University of Paris and the Sorbonne, undertaken by Brown, reveals that "throughout history, almost without exception, the expressed or clearly implicit goals of colleges and universities have been to have an impact on students in ways more extensive than passing on facts, specific skills, or intellectual capacities (10.1)." The American Council on Education (ACE) first in 1938, and again in 1949, presented the following four basic assumptions in its Student Personnel Point of View statement: (1) the individual student must be considered as a whole; (2) each student is a unique person and must be treated as such; (3) the total environment of the student is educational and must be used to achieve his/her full development; and (4) the major responsibility for a student's personal and social development rests with the student and his/her personal resources (10.2).

Although there are those who contend that what the individual student does of a personal, social, or political nature is not really the concern of an educational institution, Mayhew implies that all kinds of nonintellectual learning should be part of a college's mission because they develop the whole student (10.3). To fulfill its mandate, an institution must act on the presumption that each student has many developmental needs which must be met in a variety of ways, both formal and informal, and that no two students have the same requirements. Further, the college does not prescribe what the student shall learn. Rather, it provides resources and opportunities and helps students use them to their best advantage.

Another influential proponent of the "whole-person" philosophy is Sanford who advocates a broad general education to help students see their productive roles in perspective, develop values which can withstand organizational pressures, and lead meaningful lives apart from their occupations (10.4). By education for individual development, Sanford means a program consciously undertaken to promote an identity based on qualities such as flexibility, creativity, open-mindedness, and responsibility. Not limited to cognitive development alone, individual development represents, in part, an educational approach concerned with the emotional and physical growth of students as well.

Accomplishing this calls for bringing all campus resources into play in order to go beyond the traditional subject matter.

Despite the broad implications of these views, the student development philosophy has usually been implemented only in separate and supplementary programs. These have come to be called "student services." Wrenn stated that "student personnel services and instructional services together form the educational program of the institution (10.5)." And Cowley maintained that "these activities appear to me . . . to be complementary to the core teaching and research functions of colleges and universities. The fact stands out clearly that the distinguishing characteristic of all members of all the groups in your fields is this: you serve students in various noncurricular ways (10.6)."

In an effort to assist students in their academic pursuits, student personnel services are organized to make meaningful experiences available and to provide necessities such as food, lodging, health care, employment, financial assistance, and recreation. Personnel services have been, and continue to be, a growing function of the university, especially during the last two or three decades. Due to increased enrollments during the 1960's and early 1970's in higher education institutions, such services have been enlarged to meet the greater number of students they serve. There have also been significant improvements in the quality of services provided. Many college students come from homes which provide numerous comforts in daily living, thus reflecting the rising standard of living. As a result, there has been a demand for universities to provide more personnel services for their students.

A four-year study, conducted by the American Council on Education from 1946 to 1950, set forth six functions of a personnel program for a college or university. They are (1) counseling services, (2) social activities, (3) financial aid, (4) psychology and educational testing, (5) housing services, and (6) research (10.7).

Williamson also lists the 17 basic elements of a student personnel program as being (10.8):

1. Process of admissions
2. Maintenance of personnel records
3. Counseling services
4. Physical and mental health services
5. Remedial services
6. Housing and food services
7. Activities programs
8. Group activities
9. Recreational activities
10. Disciplinary counseling
11. Financial aid program
12. Student employment service
13. Placement service for graduates
14. Counseling services for foreign students
15. Religious activities program
16. Counseling services for married students
17. Evaluation of student personnel services

Selected elements of a student personnel program are discussed in the following sections. The discussion will be from the total university view and will include the college of agriculture and/or the teacher education in agriculture unit in some elements.

Financial Assistance

The administration of scholarships, loans, grants, work-study programs, and assistantships is an important function of student services in colleges or universities. Some students, because of limited resources, find it difficult to finance a university education. Universities provide financial assistance, to the extent possible, to such students. An attempt is made to provide each student with a balanced package of financial assistance comprised of grants, loans, and work assistance. Students with relatively small needs typically will receive loans. Students with substantial needs usually receive a combination of grants, loans, and/or work assistance.

The student services staff in the student financial aid unit should ensure that financial aid information is available to interested students. As of 1980, there were five basic types of financial aid: Basic Education Opportunity Grant (BEOG), National Direct Student Loan (NDSL), Supplemental Education Opportunity Grant (SEOG), College Work Study Program (CWSP), and Guaranteed Student Loan Program (GSLP) or Federally Insured Student Loans (FISL).

In addition to these types of basic aid, college and universities have numerous scholarships available for students. In general, these scholarships are supported by friends and alumni of the university and business and industry. Scholarships are usually awarded based on the students' academic and leadership abilities, although criteria for awarding some scholarships do include financial need. Within the colleges of agriculture, scholarships for students in agriculture are usually supported by alumni and the agricultural industry. A considerable number of these scholarships are earmarked; that is, they are limited to students in a particular major, grade classification, geographic area, or other specified categories. Many colleges of agriculture have a student scholarship banquet each year to honor the recipients and to recognize the sponsors.

At the graduate level, many agricultural education programs have assistantships for full-time students who are working toward a master's or doctor's degree. These assistantships may be either teaching or research assistantships. Current practice is for The Agricultural Education Magazine to include a report on the assistantships available for the current year (usually in the January or February issue).

Counseling Services/Program

Universities have a staff of trained counselors whose primary function is helping students work through educational, career, and personal problems. Some typical student concerns are ineffective study skills and habits, inadequate reading skills, difficulty in adjusting to the college environment, inability to engage in career planning, difficulty in choosing an academic major, inadequate interpersonal relations, feelings of depression, emotional insecurity, and specific problems of a personal nature. Interviews are confidential and voluntary. Tests of intelligence, aptitude, achievement, personality, vocational interest, study skills, and others are used.

Many institutions offer special classes/programs in remedial reading and learning skill acquisition on a non-credit basis. Individual and group counseling settings are available to students.

Housing/Dining

Many educators believe that students develop a sense of identity, security, and belonging, as well as enthusiasm and stimulation for learning outside the classroom. Many educators recommend university housing as the best way to meet people, make friends, and generally become involved in campus life. The residential setting should contribute to the students' education by helping to ensure a pleasant environment for living.

A high percentage of students live in university housing facilities. The percent of entering freshmen living in university housing is generally high and usually decreases each year as students advance toward graduation. Most universities offer dormitory facilities and apartments for students. Apartments are provided for graduate and/or married students. Also, some universities operate mobile home parks for student housing. In addition, most universities assist students in locating off-campus housing by maintaining a list of available rooms, apartments, and houses. Colleges of agriculture often provide housing on university farms and other facilities for many students majoring in agriculture. In most cases, students work on the farms and in the greenhouses, laboratories, etc., in return for their housing.

A significant number of students who live in university housing also utilize university dining halls for their meals. In fact, many of the major universities feed more individuals each year than any other food business in their states. Students usually have several different options as to the meal plan that they purchase.

Health Service

Institutions of higher education almost universally provide health services to students. Generally, a student health office or service is provided. Often the registration/tuition fee includes a health fee which entitles students to the services provided by the health program. In other cases, the health fee may be optional. Those students not choosing the program may use the services on a fee-for-service basis.

Intramural Sports

Even though most universities and colleges are best known for their intercollegiate sports teams, many students spend considerable leisure time in physical activity through the university intramural sports program. Intramural flag football, volleyball, softball, and basketball involve large numbers of college students. These activities are usually supervised by a director of campus recreation. Many clubs and organizations in the college department of agriculture have teams and participate in these intramural sport activities. Collegiate FFA chapters and agricultural education clubs/societies usually have students participating in these sports as one of their activities.

Student and University Government

Opportunities exist for students to play a role in the affairs of their institutions. Student government serves as the official voice of students within a university community. Generally, student government, which is charged with the responsibility of recommending policy considerations in the areas of academic affairs and student life, provides numerous student services. Student government is modeled after our national government, having legislative, executive, and judicial branches.

Colleges and departments of agriculture make use of student government/ student council models. These councils (composed of representative members) play a key role in advising administration and faculty on student concerns and desires. Another function is to provide leadership and assistance in conducting many college activities for students.

Professional Student Organizations

Students preparing to become teachers of agriculture have several organizations in which they can become active members. Some of the more important ones are (1) Alpha Tau Alpha chapters, (2) collegiate FFA chapters, (3) local agricultural education clubs/societies, and (4) local associations of the Student National Education Association. Each of these organizations strives to help prepare students for their role as agricultural educators. A brief description of each is presented in the sections that follow.

Alpha Tau Alpha

Alpha Tau Alpha (ATA) is a national professional honorary agricultural education fraternity that plays an important role in the preparation of students who plan to teach vocational agriculture. National in scope, it originated in 1921 at the University of Illinois. A brief history of ATA's origin has been given by Dr. Aretas W. Nolen (10.9), who was professor and head of agricultural education at the University of Illinois in 1921.

> It was at the end of a busy day one winter evening in 1921, just as I was about to close my office, when three young men from one of my classes in agricultural education diffidently stepped in, and asked if they might explain a proposition that was upon their minds. We sat down together, and after an hour or more of earnest conversation and dreaming, a new fraternity for prospective teachers of vocational agriculture was launched. We decided that a group of men, unified by such a high purpose as to become good teachers of agriculture, and to serve a nationwide cause so honorable, had good

reasons for a formal organization. Why not band together in a professional fraternity, and enjoy the benefits which such fraternities have brought to other groups, with no more worthy motives or causes for organization than teachers of vocational agriculture have?

The stated purposes of ATA are (1) to develop a professional spirit in the teaching of agriculture, (2) to help train teachers of agriculture who will be leaders in their communities, and (3) to foster a fraternal spirit among students in teacher education in vocational agriculture.

The Alpha Tau Alpha Official Manual presents a detailed history and information about the organization. The manual reveals 41 chapters, with 34 listed as active chapters (10.10).

In November 1979, in Kansas City, Alpha Tau Alpha held its thirty-ninth annual conclave. Individual chapters select a delegate and an alternate to send to each annual conclave. National officers of president, first vice-president, second vice-president, and secretary-treasurer are elected for four-year terms. For almost 30 years, the national conclave has been held at the same time as the national FFA convention. The program is coordinated with the National Conference of Student Teachers in Agricultural Education as well as with the national FFA convention.

Collegiate FFA

Membership in collegiate chapters of the FFA is available to students in many colleges and universities. An undated directory of teacher education groups in agricultural education, received in March 1979 from the national FFA office, indicates that there are 29 collegiate chapters organized in colleges and universities in 21 states (10.11).

Collegiate chapters initially existed for the purpose of helping prospective teachers of agriculture prepare for their role as advisors of local FFA chapters. However, the membership and purposes of collegiate chapters has been broadened in the past decade. According to the 1981 Official FFA Manual, the delegates at the 1970 national FFA convention changed the constitution to state that "collegiate chapters may be established in two or four year institutions where agriculture courses are taught. Membership shall be opened to students enrolled in agricultural courses, or who are pursuing career objectives in the industry of agriculture (10.12)."

Collegiate chapters are chartered by and operated under the authority of the state FFA association. The 1981 Official FFA Manual (10.13) states that "collegiate chapter members shall pay State and National FFA dues." Usually a faculty member in agricultural education serves as advisor of the collegiate chapter.

Agricultural Education Clubs/Societies

Agricultural education clubs and/or agricultural education societies are locally organized groups designed for majors in agricultural education. A directory of teacher education groups in agricultural education lists agricultural education clubs and/or societies in some 23 colleges and universities in 15 different states (10.14). Generally, the purpose of these organizations is to help recruit and prepare students to be effective teachers of agriculture and advisors of the FFA. Since these organizations are local in nature, there is no standard pattern of operation. However, many of these clubs have developed emblems, membership pins, and banners, and they carry out a planned program of activities.

Student National Education Association

The Student National Education Association (SNEA) was organized in 1937 as a department of the National Education Association (NEA). In 1976, the SNEA incorporated as a semi-independent association. Student NEA's status is currently like any state teachers association. Its structure is similar to that of the National Education Association. It has two full-time student officers, an executive director, professional staff and support staff in the NEA headquarters in Washington, D.C.

The SNEA claims to be the largest student membership organization in the world. Data in a letter from the SNEA executive director (10.15) indicate there were approximately 30,000 members in 1981-82.

The purposes of the SNEA, as stated in its Governance Documents, are as follows (10.16):

> The purposes of the Association shall be to develop in prospective educators an understanding of the education profession, to provide for a united student voice in matters affecting their profession, to influence the conditions under which prospective educators are prepared, to advance the interests and welfare of students preparing for a career in education, to forward the aim of quality education, to promote and protect human and civil rights, and to stimulate the highest ideals of professional ethics, standards, and attitudes.

National Student Teacher Conference

In 1951, the agricultural education profession began the National Conferences of Student Teachers in Agricultural Education. The conference has always been held in Kansas City, in conjunction with the national Alpha Tau Alpha conclave and the national Future Teachers of America convention.

Since its beginning in 1951, the National Student Teacher Conference has grown and expanded in scope and student participation. Lee indicates that approximately 275 students and 50 faculty members from teacher education departments in the United States attended the conference in 1978 (10.17).

Departments of agricultural teacher education in different universities assume responsibility for developing and conducting the conference program, which is designed to stimulate the professional development of preservice teacher education students. The National Vocational Agriculture Teachers Association (NVATA) sponsors a reception and presents a program on NVATA activities. The conference usually includes a presentation by a well-known leader in agricultural education. The program includes presentations by successful teachers and discussion groups led by student teachers to help keep the conference practical. In addition, for several years Farm Land Industries has sponsored a student teachers dinner, which is attended by about 300. In 1978, a breakfast sponsored by Ring Around Products was added to the program.

Membership in Professional Agricultural Organizations

Student membership is available to agricultural education students in both the American Vocational Association and the National Vocational Agriculture Teachers Association. In addition to being members of the national organization, many students are members of their state organizations.

Student membership in NVATA has grown tremendously in recent years. Stenzel reports student membership began in NVATA in 1959-60 with 15 members, reached 796 in the 1974-75 year, and grew to 1,448 in 1977-78 (10.18). At the national student teachers conference, the NVATA recognizes those institutions which report the attainment of 100 percent membership.

Through NVATA membership, students become acquainted with their professional organization and learn to understand the responsibilities of membership in the vocational agriculture teaching profession.

Leroy reports a total student membership in the AVA of 2,558 for 1978-79 (10.19). Of this total, 200 are from agriculture. Student members in the AVA receive the AVA newspaper, Update (nine issues), and the AVA journal, VocEd.

Advising of Students

Generally, students in agricultural education have faculty members in agricultural education as their advisors. The primary responsibility of advisors is to help students plan their academic programs--which involves identifying specific courses needed and the sequence in which they will be taken. Good advisors attempt to help students select routes which will give the maximum opportunity to complete their programs successfully. They must help students select schedules that provide for graduation and certification requirements, employment, and family responsibilities. They should provide students with facts and with alternatives regarding their academic programs; however, students should make the ultimate decisions.

Even though an advisor's primary responsibility is curricular selection, a teacher educator should also guide advisees in preparing for the teaching role in other ways. These include helping the students to identify and correct deficiencies, gain experience in working with agriculture teachers and other students, and participate in leadership development organizations. It is important that faculty advisors be interested in working with students and willing to take the time to help them with their problems.

Agricultural education faculty members usually have limited preparation in providing counseling for advisees. Therefore, advisors should refer advisees to other sources of assistance when a problem is beyond their ability to help.

Placements of Graduates

More and more universities are becoming concerned with the placement of their graduates. One hundred years ago little effort was made in a formalized manner to place teacher education graduates. Currently, almost all universities maintain a placement service office for students, and many offer a teacher placement office as well. Most colleges of agriculture provide considerable assistance to graduates in finding employment. In increasing numbers, they are assigning personnel specific responsibilities for placing graduates. In addition, many colleges now conduct career days on which representatives from agricultural business and industry can explain the employment opportunities available to college graduates.

In agricultural education, considerable help is given to students in finding employment as teachers of agriculture. Teacher educators in agriculture generally accept the placement of their graduates as a responsibility of their position. In many states, teacher educators work closely with state and district supervisors to help provide local school districts with the best possible agriculture teachers.

Most departments of agricultural education conduct seminars for their students on preparing a résumé, applying for a position, and participating in the job interview. Many students have reported that the information and practice secured in these sessions were the deciding factors in their getting a job.

Follow-Up of Graduates

Follow-up of graduates has been a part of most teacher education programs in agricultural education for many years. With the increasing emphasis on accountability, placement is likely to receive even more attention in the future.

Many teacher education programs are now providing courses for beginning teachers of agriculture. These courses, offered throughout the year, usually meet on a weekend schedule. The topics discussed range over a wide subject area and are geared to the specific needs and problems encountered by beginning teachers. Often, these courses involve the entire department faculty as well as members of the state supervisory staff. A part of the course involves visits to the beginning teachers' schools. Teacher educators will spend two or more days during the year with the teachers in the schools. These visits are intended to provide encouragement and assistance to the teachers. Another real benefit is in helping the teacher educators evaluate the products of their educational programs and to help them keep up-to-date about teaching vocational agriculture.

Another means of follow-up utilized by colleges/departments of agriculture is a follow-up survey. Many departments survey their graduates on a regular basis (usually every five years) to identify placement and occupational success and to obtain feedback to improve the teacher education program.

Colleges and universities are giving increasing emphasis to developing strong alumni associations and relationships. Colleges of agriculture in growing numbers are designating full-time personnel to direct alumni organizations.

Summary

Student personnel services make an important contribution to the preparation of teachers of agriculture. Some services, such as housing/dining and medical care are basic to an environment that is conducive to learning. Other services, such as student organizations, provide the opportunity for the development of competencies, such as leadership, that are necessary for success in teaching agriculture.

Student placement and follow-up provide the job skills and information needed by graduates to secure employment, while at the same time facilitating recruitment by local school districts. Additionally, follow-up data on graduates are necessary for meaningful evaluation of teacher education programs.

CHAPTER 11

Inservice Education for Teachers of Agriculture

Glen C. Shinn
MISSISSIPPI STATE UNIVERSITY

Joe P. Bail
CORNELL UNIVERSITY

Inservice education, in its broadest dimension, can be defined as any professional activity which purports to upgrade the performance of a teacher. It may consist of a one-hour presentation by an expert in a specific area, or it may be a credit course which meets regularly throughout the year. Within this wide range are many types of activities including conferences, travel seminars, individual study, workshops, and short courses.

Good defines "inservice education" in the Dictionary of Education as "efforts to promote by appropriate means the professional growth and development of workers while on the job; . . . includes planned and organized efforts to improve the knowledge, skill, and attitudes of instructional staff members to make them more effective on the job . . . (11.1)."

Note that this definition does not put limits on the activity as to credit or non-credit, length of time, sources, or methods of delivery. Because of changes in technology, practices, and philosophy, the scope of such activities should be very broad. Nor do all such activities have immediate results in changing teachers' teaching procedures. Attitudes may be changed without teachers altering their basic teaching methodology.

However, if an inservice activity is to be meaningful and useful to the participants, it must include at least three phases: (1) a pre-assessment of where the individual teacher is now in regard to the activity; (2) a specific plan for meeting the educational objectives outlined; and (3) a post-assessment of the degree of success of the activity in terms of the individual teacher's level of development at the conclusion of the activity. Adoption and application of newly acquired skills take place later and should be evaluated at that time.

Role of Inservice Education

The expectations for inservice education have increased in both the professional and the public realms. The expansion of knowledge, development of new technology, and improvements in mass communications have contributed to this increase. The need to keep up-to-date is present in all professions. The teacher of agriculture must be current in both agricultural sciences and educational developments. The technology of agriculture and education is changing. The good management practices of today are rapidly becoming obsolete. The effective teacher must be current and knowledgeable about new developments. At one time education was the only profession with structured inservice requirements. Teachers moved from annual or provisional certification to permanent certificates. Today, many skilled and professional workers are required to participate systematically in inservice programs in order to maintain their credentials and technical competence.

Even the best preservice program is limited in scope and application. Undergraduate students have limited experiences to use as a reference for synthesizing and evaluating methods or procedures. Program planning may be an abstract concept until the student becomes a teacher. Merely requiring participation in such activity does not in itself assure competence. Performance must be accurately evaluated to determine if the desired changes have come about. The Pennsylvania State Plan for Vocational Education is an excellent example of a comprehensive system for inservice training of teachers of agriculture (11.2).

Target Groups of Inservice Education

It is no longer acceptable to speak of teachers of agriculture as a monolithic group. Specializations in subject matter, variations in preservice training, and personal socio-economic factors lead to much diversity. Target groups must then be those with commonality, as determined by a sophisticated needs assessment. The New York model, carried out in cooperation with the teacher's organization and carefully articulated by the teacher trainers and supervisors in the state, is one such example (11.3). Data gathered herein and sorted on the basis of key descriptors should prove useful in planning an inservice program.

Some examples of descriptors which may be used to identify specific needs of target groups are as follows:

- Area of program specialization
- Preparation of teacher

  Traditional (entry from college)
  Non-traditional (entry from industry/business)

- Geographical location

  Rural community
  Suburban community
  Urban community

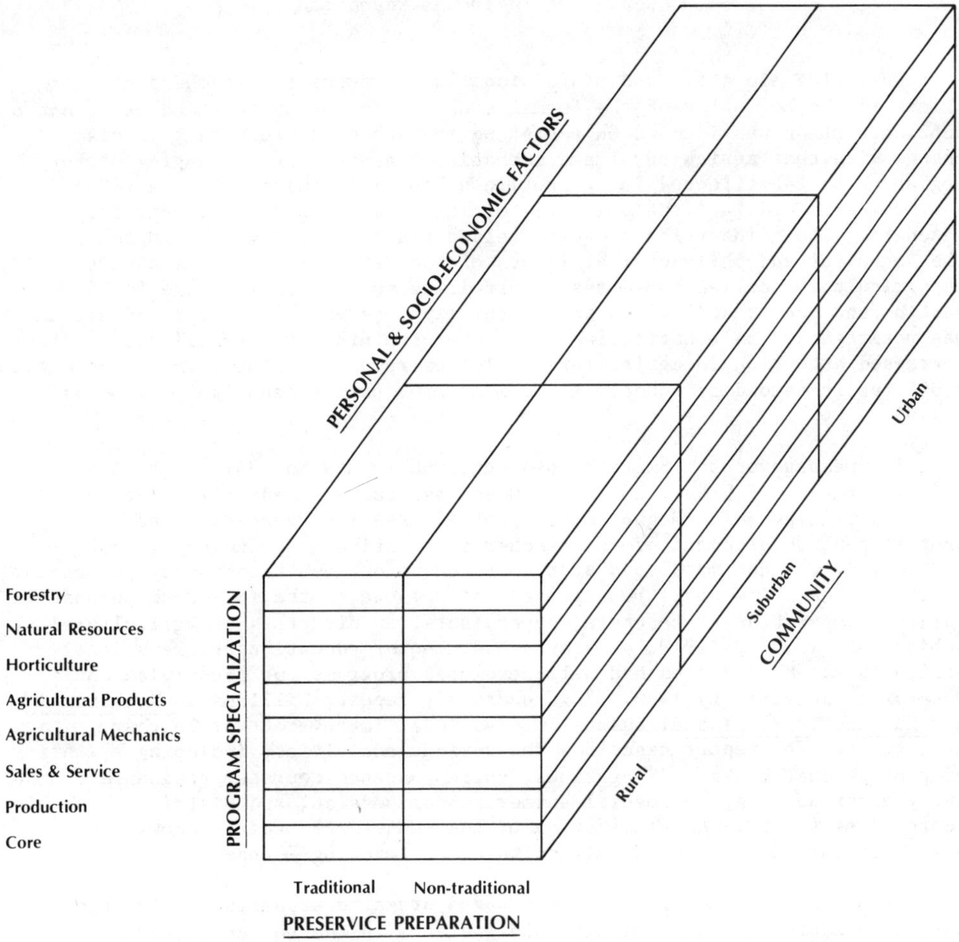

FIGURE 10.1--Factors affecting the inservice needs of teachers of agriculture.

In the past, efforts were largely directed to two groups of teachers: beginning teachers and experienced teachers. While this dichotomy has merit, it does not assess the needs of individuals. While it is necessary because of financial and other constraints to organize teachers into groups, individual needs should not be lost in the process.

Finally, professional (pedagogical) and technical needs are not necessarily the same for all individuals. This is especially true with regard to teaching skills, since greater variation in education and background may exist. Presumably, the beginning teacher should have the basic technical knowledges and skills. However, this frequently is not the case with reference to pedagogical training.

History of Inservice Education

Inservice education and professional development for teachers had its origin in the beginning of the formal school. Following the Land-Grant Act of 1862, attention was focused on assisting the school to respond to social change with the "agricultural and mechanical" aspects of a changing society beginning to be reflected in the public school curriculum. By the 1880's, many colleges and institutions were holding summer courses designed for teachers. These inservice programs helped teachers deal with changes in both the technical and philosophical issues of the day. True reports that by 1910, 46 agricultural colleges had teacher-training work in agriculture (11.4). Martin concludes that "this work was chiefly inservice in nature and evidently was not regarded as constituting a 'teacher-training program (11.5).'" This increased attention in agricultural sciences resulted in programs in more than 3,000 public secondary schools as well as many normal schools by 1915-1916 (11.6).

The passage of the Smith-Hughes Act, Public Law No. 347, by the 64th U.S. Congress on February 23, 1917, moved agricultural education into a new era (11.7). The Smith-Hughes Act supported inservice education and professional development of the teacher of vocational agriculture. As written, ". . . not more than sixty per centum or less than twenty per centum of the monies . . . shall be expended for any one of the following purposes: For the preparation of teachers, supervisors, or directors of agricultural subjects. . . ." By 1919, 40 states had teacher education programs in agriculture. Many states had well-developed programs for preservice and inservice education by 1930. In a quarterly report, Williams in Vocational Agriculture in Florida discusses "Professional Improvement of Teacher In-service." The report describes the various activities, including a listing of professional books and magazines, correspondence courses, residence courses every other Saturday, a special summer school session, and district and state conferences (11.8). An examination of the conference program shows time allocated for both technical agriculture and teaching methods.

In the 1940's and 1950's, there was a dramatic expansion of vocational agriculture programs and a remarkable increase in the number of teachers. Programs in Missouri, as an example, grew from 175 in 1945-46 to 262 in 1959-60. Along with this increase in teachers, the state supervisory staff and teacher education faculty also increased.

Itinerant Teacher Trainers

The inservice education function was usually shared by the university and the state departments of public instruction. The individuals with inservice responsibility were often designated as "itinerant teacher trainers." The exact duties of these teachers trainers varied among states. Generally, they were considered to have full responsibility for teacher improvement and program development but none of the administrative or fiscal responsibilities of the supervisors. Traveling throughout their states, itinerant teacher trainers conducted small group meetings, workshops, and short courses. Much of the time was spent on individual instruction in local programs. The job title "itinerant teacher trainer" had disappeared by the late 1950's and early 1960's, but this concept has remained important to program and professional development. Today, many teacher educators maintain inservice education responsibilities, and in many states "program consultants" employed by state departments of public instruction conduct a variety of non-credit professional improvement activities.

As inservice and professional development activities matured, credit courses moved from the college campuses to resident centers, which has resulted in "statewide" campuses. The concept of inservice education has broadened to include credit courses, non-credit courses, workshops, short courses, conferences, seminars, and other field-based experiences. Today, several states are heavily committed to inservice activities for teachers.

In 1979, an ad hoc committee on inservice education was commissioned by the American Association of Teacher Educators in Agriculture. In a review of national inservice activities, Shinn reported that in fiscal year 1978, 94 percent of the vocational agriculture teachers in Missouri participated in a statewide program development workshop (11.9). Forty-two workshops also were conducted by The Pennsylvania State University in fiscal year 1978, with 834 teachers and extension agents in attendance. This AATEA report concluded that "inservice education is an important activity if agricultural educators are to remain as an informational source and change agent in an era of electronic communications and 800- telephone numbers (11.10)."

Need for Inservice Education

In an age of increased program accountability and keen competition for scarce resources, the tasks of needs assessment and program documentation are a priority. The need for inservice education can be demonstrated by rapidly changing agricultural technology, numerous internal changes within the community and agricultural industry, and the increasing number of teachers who enter the profession each year. This documentation should be made by a committee of teachers from the professional teachers organization which represents geographic areas and/or program specialities. This joint committee should also include representatives from the teacher education faculty and the state supervisory staff. As a primary responsibility, the teacher representatives should poll their subgroup to determine the needs for credit and non-credit courses, workshops, and other professional development activities. These needs can be collectively developed by the committee into a one-year and a three-year plan to make maximum use of financial resources and professional faculty and to provide program continuity. Another primary responsibility of the joint committee is to serve as a "third-party audit" for program evaluation. Many special inservice projects will require evaluation from both the participants and members of a third-party group. These individuals are an important part of the evaluation process.

Planning and Organizing Inservice Programs

A variety of educational experiences is necessary if the profession is to respond to inservice needs. Reliance on any one activity, such as college courses or summer conferences, is unlikely to meet the needs. While each may make a significant contribution, the sum of the whole is necessary to provide maximum professional development.

Organizational Prototype

As a result of leadership from the Missouri Vocational Agriculture Teachers Association (MVATA), the Missouri Department of Elementary and Secondary Education (MDESE), and the University of Missouri - Columbia (UMC), a joint committee for inservice education and professional development was established in 1975. The membership was initially composed of one vocational agriculture teacher from each of the 13 geographic areas within the state, two MDESE representatives who were district supervisors for the agricultural education section, and two UMC representatives who were involved in the delivery of inservice courses in agricultural education. A presiding officer and a secretary were elected from within the group.

A three-year rotation was established for committee members, with each of the four districts in the state having a member in each of the one-year, two-year, and three-year terms. After the first year, each representative was elected by the area teachers group for a three-year term. The presiding officer and secretary both serve for one-year terms. The MDESE and UMC representatives serve ex-officio terms, which are consistent with their job appointments.

The committee had four primary responsibilities in the development of an inservice program.

1. Determine the inservice needs of teachers within the geographic areas.

2. Coordinate long-range planning and assist in providing data for priority determination.

3. Monitor all inservice activities.

4. Evaluate all activities in the geographic area.

The committee meets prior to the summer teachers conference and reviews the program of activities. During this meeting, a review of all inservice and professional development activities is complete. The presiding officer and secretary assume their duties. Recommendations are developed for special projects, and the annual plan is initiated.

The second committee meeting is held during the following winter MVATA executive officers meeting. The primary agenda includes the modification and extension of the three-year plan. Ideas are developed for summer non-credit activities, and resource persons are identified.

Elements of this prototype that have general applicability for organization and planning of inservice programs are cooperative efforts by the parties involved, systematic needs assessment, one-year and three-year projected planning, consideration of the needs of individual groups, and planned and continuing evaluation.

Inservice Delivery Systems

As in selecting methods for effective teaching, one should determine the instructional objectives prior to determining the system for delivery. Credit courses or short courses will require a different delivery system than a conference will.

A historical review of the literature describes several systems of delivery. Credit courses were ". . . offered every other Saturday morning during the Fall Semester (11.11)." Many universities were reluctant to offer graduate-level credit courses away from the campus. Oklahoma was a pioneer in the early 1940's in developing a credit course for beginning teachers. Initially, this course was offered on the Oklahoma A&M College campus, then in district meetings, and finally on an individual basis. Topics included practical program planning, course development, leadership, and program management. The sessions were conducted jointly by teacher education faculty and state supervisory staff. Gradually, other states initiated credit courses in residence centers and in out-state locations. Today, some institutions do not distinguish between "resident" and "non-resident" courses for advanced degrees.

Some states use the professional improvement (PI) system. A PI-group represents a group of teachers in need of professional improvement in a geographic area. Oklahoma subdivides the state into 22 PI-groups. These meet on a regular basis and deal with a broad range of topics over the year. Usually the district supervisor meets with the group.

Other delivery systems that are used include special camps, or retreats. Arkansas holds a one-week session at the state FFA camp. The session includes a full range of professional improvement activities for all vocational agriculture teachers. Teachers develop technical and professional competence, as well as have the opportunity to engage in social activities and make new friends. Purdue University has developed a delivery system for off-campus credit courses which utilizes the Indiana educational TV network. The initial course meeting is organized by the instructor at the off-campus site. Other class sessions are taped and carried by the network into satellite centers. Following the session, the instructor can discuss specific problems or questions using a telephone connection. Moore reports on the use of audiotapes to interact with beginning teachers (11.12).

Tomorrow's delivery systems may be very different from those used today. For example, commercial companies have developed the technology for conference telephones and electronic blackboards. These alternative systems will continue to play an increasingly important role in the inservice education program.

Additionally, formal acknowledgement of participation in inservice activities must be made. This may include certificates, continuing education units, and other types of evidence of participation. Such acknowledgement will be useful in determining assignment of local inservice credit by boards of education in local schools.

Several of the most commonly used inservice delivery systems are described in more detail as follows.

Conferences, Workshops, and Short Courses. A conference is described by Good as "a meeting of two or more persons of common interest who come together primarily for consultation, discussion, and interchange of opinions and ideas (11.13)." Teacher organizations have increasingly assumed responsibilities of most state conferences and even national conferences. To the extent that good leadership and follow-up activities are practiced, this has been an effective inservice mode. Careful attention must be given to planning and execution to make such conferences effective. The selection of outstanding resource persons with clear-cut instructional objectives is a primary ingredient of success. It is also important that key representatives of all cooperating or sponsoring agencies be involved in the planning process.

Workshops in agricultural education are conducted as specialized meetings "in which persons with common interests and problems meet with appropriate specialists to acquire necessary information and develop solutions through group study (11.14)." The workshop format implies active involvement by participants. Workshops in agricultural education are conducted as specialized meetings on specific topics. They are usually one day or less in duration and do not carry college credit.

A short course is defined as "a form of class or correspondence course for adults, less extended and formalized than regular courses offered by colleges or universities (11.15)." These courses may be offered with or without college credit. Short courses usually meet for an extended session, three to four hours, and for two to four sessions. This design is usually arranged off campus and deals with a specific problem or activity. Several agricultural mechanics activities, such as small engine overhaul, diesel maintenance, and electric motor usage, lend themselves well to this delivery system. Regardless of whether the activity is a conference, workshop, or short course, the basic principles apply. The title, however, should be consistent with the format and goals of the activity.

Other inservice activities include seminars on wheels, retreats, open-ended meetings, and individual reports of successful activities in the classroom, laboratory, or other settings. The sharing of instructional resources should not be overlooked.

The sharing of ideas, whether in a formal or an informal way, is a key to successful professional improvement. Joint planning and execution, without regard to recognition for those involved, is essential.

Evaluating Inservice Programs

Historically, evaluation of inservice has been a post-program description of activities which was used more for compliance than for evaluation. The evaluation of inservice and professional development activities should (1) determine the overall effectiveness of the program and (2) provide formative data for the modification and improvement of future activities. Much of today's evaluation technology is useful in determining the quality and effectiveness of inservice activities. The evaluation process should include data related to the reaction of the participants, the learning outcomes, and the resulting changes in behavior. The procedure should be planned and congruent with the goals and objectives of the inservice or professional development activity.

Formative Evaluation

Evaluation begins with a needs assessment. A teacher survey is important to identify priority inservice needs of teachers. These data should be analyzed to establish the type and level of delivery. If a general need is prevalent and the subject is best delivered to small groups, the activity should be planned as a statewide workshop or as a short course. If the subject is a high priority for specialized teachers but requires sophisticated equipment, the activity may be best delivered during a state conference workshop or an on-campus credit course. This survey may be conducted during a group meeting or as a mail-out instrument. In any event, concerted efforts should be made to get teachers to think in terms of their "real" needs. Needs assessment data should feed into the state plan and financial planning. Ideally, future state and federal legislation dealing with professional development would use this unit planning data to establish or modify training priorities.

Once the priorities are identified, goals and objectives should be developed for each inservice activity.

Unobtrusive formative evaluation techniques should be used as an on-going procedure to estimate the effectiveness of each activity. Teacher interest, participation, and subtle gestures provide a basis for internal evaluation and modification of methods, or techniques.

Summative Evaluation

Summative evaluation of inservice and professional development activities may be classed into three broad categories:

1. Reaction of participants
2. Learning outcomes
3. Behavioral changes

<u>Reaction of Participants</u>. The American Society for Training and Development guidelines for constructing instruments provide data which reflect the reaction of teachers who were involved in the activity (11.16). The instruments should provide quantifiable data as to how each participant felt about the short course and a standardized comparison among inservice activities. Regardless of the format, the instrument should allow teacher input on an anonymous basis. Such an instrument allows a quick summary of the activity. It may also be used as formative data when the activity is repeated in several locations. Following the inservice series, data on each activity may be summarized by using frequency distributions and then combined to provide an overall reaction summary.

<u>Learning Outcomes</u>. The second type of summative evaluation should measure learning outcomes. Every activity should be measured as to the amount of learning that takes place. Evaluation may be by demonstration, pencil-and-paper test, or a combination of both. These test data should be used to help determine the effectiveness of the inservice activity.

Behavioral Changes. A third category of summative evaluation aids in determining changes in practices or attitudes. This should be a longitudinal survey which reports behavioral change reflecting the use of the information. If a specific technical area, for example, gas welding and cutting, was a priority need and a significant number of teachers participated in a series of short courses, then higher work quality in the program and possibly more time allocated to teaching the subject should be expected. This change should be evident through long-term replicated observations.

Financing Inservice Education

Programs of inservice education will likely be financed from a variety of sources. Increased demands and needs have led many agencies to become involved. Chief among these are state departments of education, which may earmark certain funds from special federal legislation or other professional development grants for inservice activities. Practices in the late 1960's and early 1970's led many teachers to expect the full support of all such activities. Not only were tuition and fees paid, but in many instances, living stipends were provided to participants. As funds began to be limited, a more diverse funding pattern began to appear. This included special grants from government, business, industry, plus individual contributions. This pattern is likely to continue.

When full funding was provided by a given agency, some problems of control emerged. Specifications by outside agencies to an educational institution were interpreted as infringing upon academic freedom. With several avenues of support available, such problems diminished. Joint support and responsibility by all involved in the programs should be the goal.

On occasion, specific programs may be proposed by an agency or organization. If such proposals are accepted, a clear memorandum of understanding should be developed prior to accepting the funds. In addition, requirements for follow-up, if any, should be clearly outlined.

Role of Supervisors in Inservice Education

Supervisors have played, and will continue to play, a major role in determining inservice programs. Supervisors, teacher educators, and teacher groups constitute the three major groups involved. Because state plans for vocational education clearly outline inservice activities, supervisors have a central role in making decisions and in implementing programs. Although actual implementation of programs may be delegated to others, their overall supervision, funding, and follow-up give creditability to the total process. In fact, if one accepts the definition of supervision as improving the learning process, the supervisor's role becomes self-evident.

Successful operation of inservice programs requires the cooperation of all parties. The state agricultural supervisory staff often works closely with other vocational area personnel in planning activities designed to serve a broader group of vocational teachers. When this happens, improved articulation of programs should result. It also frequently aids administrators at local levels to understand better the various facets of an occupational education curriculum, such as youth organization, work experience, and shop safety management.

State plans for vocational education include inservice programs in both annual and long-range efforts. The responsibility for seeing that programs meet the basic requirements of the state plan, including provisions for special funding, rests with the chief state supervisor. Generally, working with state institutions or private agencies, he/she outlines a detailed proposal to fulfill all requirements. Vists and direct contact with inservice participants are generally a part of the endeavor. Finally, follow-up and evaluation of the outcomes, such as the resulting changes in teacher behavior, are done by the supervisory staff.

Most state supervisory staffs designate major committees comprised of teachers, industry representatives, and others involved in the delivery process to plan, implement, and evaluate the program. Some basic guidelines for such a committee to consider are the following (11.17):

1. The inservice program focuses on improving teaching effectiveness.
2. The inservice program will emphasize both technical and pedagogical areas.
3. The inservice program is performance based, i.e., roles are derived and specified in behavioral terms, levels of mastery to be achieved are specific, and performance is the prime evidence of mastery.
4. Inservice programs will encompass both felt and unfelt needs.
5. Inservice programs must be adequately supported at local and state levels.
6. Inservice programs clearly delineate responsibility of all parties.
7. Inservice programs meet the criteria specified if special funding is accepted.

Role of Teacher Educators

To maximize the outcomes of inservice and professional development activities, teacher educators must be actively involved. In an age of specialized agriculture, every teacher educator should identify an area within the profession and be responsible for bringing the new technology of this area to the practicing teacher. Historically, itinerant teacher trainers assumed this responsibility, and later those with dual appointments taught technical inservice courses. Today, there is less evidence of an active delivery system for inservice and professional development activities. In an ad hoc report, Shinn states that in a survey of 25 states, the modal number of workshops, short courses, and conferences for vocational agriculture teachers was zero (11.18). According to these data, teacher education programs in a majority of states have not met the responsibility to deliver inservice activities to teachers.

Because of length of service and changes in technology, vocational agriculture teachers must be provided with continuing inservice education. Teacher educators, because of organizational patterns, are in a position to assist and coordinate in determining the needs of the profession. Also, teacher education has the advantage of offering credit for off-campus courses. Teacher educators should assist in workshops dealing with curriculum development, methodology, and technical subject matter.

Teacher education units should organize to deliver non-credit as well as credit activities. Many state certification agencies require continuing education units (CEU's) to maintain teaching certificates. If these CEU's are to have any effect on teacher competency and program quality, they must be directed to the needs of the individual as well as to those of the profession.

An accounting system should be developed which will credit teacher educators for providing inservice education. Teacher education, in turn, should build the inservice function into the regular work load and reward system of the faculty. When "off campus" is viewed as an overload, few teacher educators can or will participate.

Role of Teacher Groups

If teaching is truly a profession, then the professionals therein should play the major role in determining their professional development and growth. In theory, this is correct. In actual practice, it may sometimes vary with the reality that teachers may not have adequate time or resources to plan such activities. True, teacher organizations are likely to be able to muster support for such activities, but it will entail much effort on their part.

Thus, teacher groups should work with supervisors and teacher trainers in planning for comprehensive inservice programs. A primary responsibility is to provide an accurate assessment of the inservice needs. Some special needs may be met entirely by teacher groups, and these groups may be assigned responsibility for such. Examples include exchanging ideas, planning contests, and exhibits, and sharing instructional resources. Specific responsibility for working with beginning teachers in a cooperative setting may also be useful. However, organized efforts involving courses, seminars, and workshops generally need services such as secretarial, duplicating, and record keeping, which can be found in a state office or in a teacher education institution.

Summary

Inservice education, in its broadest context, includes all planned professional activities designed to improve the technical and pedagogical competence of teachers of agriculture. In an era of changing technology and program accountability, educators should be able to quantify both the needs and benefits of inservice activities. Using programmatic evaluation data, educators can best conduct this planning and documentation.

As a profession, agricultural education has had a rich history in developing and conducting inservice education programs. However, in 1979, although several states were conducting an outstanding inservice education program, 25 states reported having no inservice conferences, workshops, or short courses for teachers of agriculture. Thus, the need for inservice in many states remains unmet.

As programs interface with a rapidly changing society, the need for a dynamic inservice education program is increased. The needs should be documented by a consortium of professionals. A needs assessment should by used to evaluate both individual and collective priority needs. A variety of delivery systems should be used to foster educational change. Evaluation of the activities should be an on-going process.

Inservice education should be directed cooperatively by those who have a vested interest. Inservice education can continue to improve if teachers of agriculture will work together toward a common goal.

CHAPTER 12

Graduate Study for Teachers of Agriculture*

John F. Thompson
THE UNIVERSITY OF WISCONSIN–MADISON

As ordinarily defined, graduate study is that formal study or education individuals pursue subsequent to having received a bachelor's degree. Graduate study usually is engaged in for the purpose of earning a degree higher than the bachelor's.

In actual practice, however, graduate study is something more than the definition implies. Graduate work is also a continuation of the study of the student's major area of interest (broadly interpreted) as an undergraduate. It is not a repetition of the undergraduate work. Rather, it is either advanced work in subjects already studied or courses related to the major interest which serve to broaden the student's background. Commenting on graduate courses rather than a graduate program, Schumann believes that "graduate courses have historically been, and will continue to be, a primary vehicle whereby professional educators upgrade their professional as well as technical skills (12.1)." This comment could be generalized to all graduate work in agricultural education. Graduate work is more specialized than undergraduate work, and it requires a strong, continuing interest and effort on the part of the individual student.

The emphasis in this chapter is on graduate work for teachers of agriculture who are seeking to earn an advanced degree with a major in agricultural education. Inservice training of a non-credit nature and graduate courses taken for credit but not for a degree are topics covered in other chapters of this book.

---

*Much of the material in this chapter is adapted from Chapter XI, "Graduate Study for Teachers of Agriculture," by A. H. Krebs, in the first edition of Teacher Education in Agriculture, 1967.

Nature of Graduate Work

What is graduate study? How does it differ from undergraduate work? Does success in college necessarily mean success in graduate school? These are not easy questions to answer. The transition from undergraduate to graduate study is normally too gradual for clear differences, except those of an extreme nature, to be identified. The differences, then, are more of degree than of kind.

The nature of graduate courses, though, is often quite different from that of undergraduate courses. Classes are generally small when compared to undergraduate classes. As one advances in graduate work, the classes become even smaller, with a typical doctoral seminar having 6 to 10 students enrolled. Papers tend to replace examinations, and students no longer take frequent quizzes. Students are expected to participate in class discussions. They must also assume much of the responsibility for their own learning.

Purposes of Graduate Study

The purposes of graduate study involve two perspectives: Why is graduate education offered? Why would teachers of agriculture seek graduate study? The first is an institutional perspective, and the second is an individual, or consumer, perspective. The reasons for each are not always the same--nor should they be.

Graduate study's primary aim is the education of scholars, teachers, and researchers--not just men and women who transmit existing knowledge but also those who actively contribute to such knowledge through research, analysis, and critique. Graduate institutions seek out persons who have high conceptual ability, as well as individuals who are fairly curious by nature (12.2).

While he was vice-president of the AVA, Warmbrod, a teacher educator, wrote: "Teachers whose goals are high quality instructional programs aggressively create and seek out ways for improving their professional and technical competence (12.3)." From an individual perspective, Attaway expresses the following:

> Realizing that if I teach and administer a complete program of vocational agriculture for all-day students, young farmers, and adult farmers in our complex society and with our rapidly changing science of agriculture, it necessitates complete preparation for the job in order to expect reasonable success. Thus, I felt it was my duty and responsibility to myself and to my future profession to plan and complete a graduate program in agricultural education (12.4).

Bail indicates that:

> If we assume that the master's degree should be the goal of most teachers, then perhaps attention should be directed to the

> reasons individuals may have for securing such a degree. These will likely include the improvement of technical and professional competencies, development of new abilities in social-personal areas, pursuing of individual interests, and personal satisfaction among others (12.5).

Bjoraker gives us still another point of view regarding the purposes for pursuing graduate study.

> Most important of all is that of developing further his ability as a teacher of agriculture. This ability then becomes translated into better programs in agricultural education, and better programs generally mean greater responsibility, greater recognition, and yes, greater pay. The graduate program can be a broadening experience including that of providing an entree into the study of new areas such as guidance or school administration. It also meets the objectives of meeting local requirements of the board of education, of additional credits or degrees in order to remain on salary schedule or to have a contract renewed (12.6).

In 1925, Broyles completed the first study of graduate work in agricultural education. In the report of his study, Broyles states that:

> The occupational objectives are, on the whole, the same for the undergraduate doing his practice teaching and the graduate returning after several years to work towards a master's degree. Each aims at being a good teacher of agriculture; each finds specific objectives in what a good teacher of agriculture must do and what he may find larger opportunities to do (12.7).

Two studies reflect the attitudes of today's student toward graduate study. Grigg, in a study of a 5 percent random sample of the freshman and senior classes of six universities in Florida, reported the following distribution of reasons for going to graduate school given by students planning to seek the master's degree (12.8).

Reasons	Distribution (%)
Financial considerations	27.5
Proficiency	20.0
Interest in subject	12.5
Professional requirements	26.4
Prestige	3.8
Altruism	9.9
Influence of others	5.0
Passive responses	3.8

LaRue, following a survey of opinions of 74 vocational agriculture teachers in Idaho during 1962-63, reported that "six factors were listed as motivating reasons for securing a master's degree. These reasons, as ranked by the teachers, were: (1) professional improvement, (2) increased financial returns, (3) prestige, (4) future promotion in profession, (5) keeping up-to-date, (6) job security (12.9)."

Allen, in an interesting article entitled "Professionalism--Who Needs It? We All Do," concludes that continuing one's education while a teacher is one step in professionalism (12.10). DeBertin reports a similar view, stressing that it keeps teachers up-to-date (12.11).

Actually, it makes little difference whether one or two specific purposes are identified as the primary forces in motivating graduate study in agricultural education so long as the work taken is closely enough related to the teacher's work that, if successfully completed, it could result in a better performance of the teaching tasks of the individual concerned.

Scope of Graduate Programs in Agricultural Education

Data taken from a survey of all agricultural education departments in 1979 depicting the magnitude of graduate programs is presented in Table 12.1.

TABLE 12.1--Number of agricultural education graduates by states and types of institutions

	Master's Students			Doctoral Students		
Type of Institution	Full-time	Part-time	Total	Full-time	Part-time	Total
Master's programs only	155	870	1,025	--	--	--
Master's and doctoral programs	206	951	1,157	89	78	167
Total	361	1,821	2,182	89	78	167

Also important is the number of degrees granted during 1978. These data are presented in Table 12.2.

TABLE 12.2--Number of agricultural education graduate degrees granted in 1978 by type of degree and type of institution

Type of Institution	Master's Degree	Doctoral Degree
Master's programs only	194	--
Master's and doctoral programs	250	46
Total	444	46

Kinds and Patterns of Graduate Programs

The kinds of graduate programs available to teachers of agriculture have, of necessity, been variations of already established graduate programs for other fields. The effect which agricultural education as a profession has had on the various kinds of graduate programs has stemmed from the patterns of courses designed especially for teachers of agriculture.

Early Graduate Courses for Teachers of Agriculture

In his survey of graduate work in agricultural education, Broyles listed many of the courses the various institutions considered for inclusion in their graduate programs for teachers of agriculture (12.12). Some of the early graduate courses in the following list are cited as a matter of historical interest and for purposes of comparison with current offerings. The courses are given as Broyles had them grouped, although the interpretive group headings are those in the first edition of this book.

Group A. Principles and Methods of Teaching Agriculture

History of Agricultural Education
Organization and Management
Principles of Method
Special Problems in Agriculture Teaching
Supervised Practice in Vocational Agriculture
Teaching Vocational Agriculture
The Agricultural High School
The Home Project
The Teaching of Animal Husbandry
The Teaching of Plant Production

Group B. Training in the Scientific Method

Scientific Method in the Study of Agricultural Education
Statistical Methods Applied to Education

Group C. Research

Advanced Problems
Research in Agricultural Education
Seminar

Group D. Administration and Supervision

Administration and Supervision of Vocational Agriculture
Organization and Management of Teacher-Training Departments
Public School Administration
The State Supervisor

Group E. Extension Techniques and Teaching Aids

Community Agricultural Extension Work
Visual Methods of Teaching Vocational Agriculture

Group F. General Education

Advanced Educational Psychology
History of Industrial and Vocational Education
Principles and Practices of Vocational Education
Problems of Rural Education
Secondary Education
Tests and Measurements

Just by reading the title, one cannot accurately determine the content of a course. However, it is apparent that the early graduate courses for teachers of agriculture were far-ranging in scope. Even the effort to develop a critical and inquiring mind through research courses was not neglected.

Current Graduate Courses for Teachers of Agriculture

Current course offerings in agricultural education carry titles quite similar to the titles of the earlier course offerings listed by Broyles. The following list of courses, obtained from a 1979 survey of all colleges and universities offering programs in agricultural education, is illustrative of what the graduate student would find today at the master's level.

Adult and Young Farmer Education
Advanced Methods of Teaching
Advanced Methods of Adult Agricultural Education
Advanced Methods of Teaching Agricultural Mechanics
Conducting Programs of Supervised Practice
Coordination of Supervised Occupational Experience Programs

Curriculum Development of Program Planning in Agricultural Education
Evaluating Programs of Agricultural Education
Experiences in Agricultural Occupations
Guidance in Agricultural Education
History and Philosophy of Vocational Education
Improved Methods of Teaching Agricultural Mechanics
Inservice Education for First-Year Teachers
Introduction to Cooperative Education
Introduction to Research
Leadership Development in Local Agricultural Education Programs
Occupational Experience in Agriculture
Policies and Programs in Agricultural Education
Post-High School Education in Agriculture
Principles of Vocational Education
Problems and Trends in Agricultural Education
Problems and Personnel in Program Development
Research Methods
Selecting Teaching Materials
Young and Adult Farmer Programs

Courses at the doctoral level, such as advanced methods, program development, and philosophy, continue some of the themes identified at the masters' level. In addition, two other themes are noted. One is the inclusion of advanced research methods and statistics. Two is the broadening of agricultural education students to include understandings of the large field of vocational education and to the foundation areas such as philosophy. Typical courses reported are:

Administration of Vocational Education
Advanced Problems in Agricultural Education
Advanced Seminar in Research Methods
College Teaching
Evaluation of Agricultural Education Programs
Evaluation of Vocational Education Programs
Independent Study
Philosophy of Education and/or Vocational Technical Education
Philosophy and Policy Making in Agricultural Education
Principles and Programs of Vocational and Practical Arts Education
Program Planning in Vocational Technical Education
Research Procedures
Research Studies in Vocational and Technical Education
Review and Synthesis of Research in Agricultural Education
Seminar in Agricultural Education
Seminar in Vocational Education
Statistics
Supervision in Vocational Education

The number of foundation courses taken by teachers of agriculture is almost unlimited. Illustrative of these offerings are the following:

Educational Psychology for Teachers
History of Educational Ideas
Library

Mental Hygiene
Rural Education Through Extension Methods
Social Foundations of Education
The Secondary School

Teachers of agriculture often minor, or major, in areas closely related to agricultural education, areas such as educational administration, educational psychology, guidance, and the various technical agriculture fields. Thus, the variety of courses which could be taken in the pursuit of an advanced degree is very broad.

Master's Level: Types of Degrees

There are several kinds of degrees which may be earned by teachers of vocational agriculture who pursue agricultural education at the master's level. While the type of master's degree is of importance to some individuals, it is usually the differences in detailed requirements which determine the particular master's degree an individual teacher will seek. A practical factor is which degree or degrees happen to be offered by the institution attended. For the most part, the kind of master's degree has no effect on the teacher's professional future. The 1979 survey indicates the following data regarding master's programs.

Master of Science. The Master of Science degree is the most frequently available degree, which is offered at approximately 35 percent of the institutions which have graduate education in agricultural education programs. The requirements for admission to candidacy for this degree usually include having earned the bachelor's degree with a major, or its equivalent, in the proposed field of study which, in this case, would be agricultural education. In many institutions, the undergraduate grade-point average must be a C+ or higher. The undergraduate grade-point average may be based on all undergraduate work or on only the last two years of work, depending on the institution. In some institutions, admission to candidacy must be preceded by successful completion (average grade of B) of a specified number of hours of graduate work. In addition to the requirements already mentioned, acceptance by the graduate department in which the work is to be done is a standard requirement.

The specific course requirements vary widely from one institution to another. For example, in one institution the only course actually required is one in the psychology of learning. In a neighboring institution, courses in statistics, research methods, library usage, foreign language proficiency (no credit), and seminar and thesis work are all mandatory. In one institution, the coursework in specialized agricultural education methodology may consist of no more than two courses, while in another institution, at least a fourth of the coursework must be in agricultural education. Requirments for coursework in general education also vary, with the range being the number of hours of one course up to a fourth of the total hours of credit needed. A range of 24 to 30 semester-hours is required for graduation. In most institutions, the major part of the planning for the master's degree is left to the student's advisor in consultation with the student.

A residence requirement for the Master of Science degree is reported at about 70 percent of the institutions. Where required, it is generally quite high, with two-thirds or more of the work to be done in residence on campus. In most cases, it is possible to meet the residence requirement by attendance at summer sessions.

Generally speaking, institutions accept only a very limited number of hours of credit toward the master's degree from another institution, and these only after a fairly rigorous examination. In several cases, however, up to one full semester of work may be transferred from an approved institution.

There is also a wide range in the length of time a student has for completing the work for the master's. A time limit of five years from the date of starting is quite common. In one case, however, this means only that the student must take at least one course every five years to keep his/her program active. Other institutions grant extensions of time upon petition by the student.

Master of Arts. The requirements for admission to candidacy for the Master of Arts degree, available at approximately 12 percent of the graduate programs in agricultural education, are generally the same as for the Master of Science degree. A bachelor's degree in the major field of study, a C+ or higher grade-point average for the undergraduate years, and acceptance by the department concerned constitute the major requirements.

As with the Master of Science, the course requirements are left largely in the hands of advisors. A total of 24 to 30 semester-hours of graduate credit are required before the degree is granted.

Transfer credits acceptable from other institutions range from approximately none to one-half of the work. Again, approximately 70 percent of the institutions require a portion of the coursework to be taken in residence, with residence being defined as courses taken on campus.

Master of Science in Education. For the most part, the bachelor's degree and a C+ or higher grade-point average are required for admission to the Master of Science in Education program, plus acceptance by the department concerned. This degree is available at approximately 14 percent of the institutions which offer agricultural education graduate programs.

Some other requirements include taking a graduate college test battery and passing an oral or a written comprehensive examination on completion of the formal coursework. In one institution, three years of teaching experience is required before the degree is granted.

The transfer of credit from other institutions is the same as for the Master of Science and the Master of Arts degrees.

Nearly 40 percent of the institutions studied do not have a residence requirement for this degree. Those that do require from 75 percent to 100 percent of the work to be taken in residence. However, in some institutions, residence is defined simply as courses in which a student is formally enrolled.

The time limits within which the degree must be earned vary from five to eight year, with extensions granted on petition of the student.

Course requirements are similar to those for the Master of Science and Master of Arts degrees.

Master of Arts in Education. No institution among those responding to this survey offers a Master of Arts in Education program.

Master of Education. The Master of Education degree description and requirements vary among institutions. All institutions considered the Master of Education degree to be a professional degree. It is second in availability, with 24 percent of the institutions offering graduate programs in agricultural education having this degree.

Admission requirements range from the applicant's being a graduate qualified teacher to his/her having a C+ or better grade-point average for the last half of the undergraduate work.

Course requirements vary from (1) one course in educational psychology; to (2) two courses in educational psychology, two courses in history and philosophy of education, one-half of the work in education, and at least one-quarter of the work in the highest level courses offered; to (3) a major in agricultural education, two minors, and a course in research. In one institution, the program includes a student teaching internship.

A final comprehensive examination is required in some cases and not in others, with the same being true for the graduate school test battery.

Nearly one-half of the institutions offering this degree do not have a residence requirement.

The time limits for earning the degree range from five years to eight years, plus extensions based on student petitions.

Master of Science in Agricultural Education. Nearly 12 percent of the institutions offer a Master of Science in Agricultural Education degree. The degree program is usually administered by the college of agriculture. Admission requirements include a Bachelor of Science degree, with a major which qualifies the candidate for a vocational agriculture teaching certificate, and a C+ or better grade-point average for the last 60 hours of undergraduate work. A graduate test battery may also be required. The degree is heavily oriented toward technical agriculture courses. Other aspects of the degree are similar to those already described.

Master of Arts in Teaching. Some interest is developing in special degrees for teachers, as exemplified by the Master of Arts in Teaching. Two institutions are known to have it available for agricultural education. The general purpose of the program is to improve pedagogical skills in a teaching field, in this case, agriculture.

Candidates must have a career interest in teaching and considerable depth of preparation in the intended teaching field. Admission to the program is considered to be of joint concern to faculty in both the education field and in the teaching area of the candidate. The program is planned by the candidate and a "special committee," which is chaired by a faculty member from the teacher preparation specialization area and which is composed of other faculty members to give adequate representation of the candidate's program. In addition to completing the planned coursework, the candidate must demonstrate teaching skill in a supervised field experience. Although a research emphasis may be built into this program, it is not necessary to do so. The degree is planned primarily for the future teacher rather than for the future researcher.

Fifth-Year Preservice Programs. Nine states offer some type of fifth-year preservice program in which the student may earn a master's degree while meeting state teaching certification requirements. These states are Alabama, Wisconsin, California, Connecticut, Illinois, Montana, New York, North Carolina, and Arizona. Differing greatly, these programs appear to have few elements common to all states. Three of the states report that their preservice programs are moving toward becoming inservice programs. Winterbourne completed a recent study of teachers who were in the fifth-year programs (12.13).

The fifth-year programs appear to be options developed by the institutions rather than options to meet state certification requirements. The main reason for the fifth-year program option is to encourage persons who have completed the work for a bachelor's degree with a technical agriculture major to qualify for teaching.

In one institution, the fifth-year program option is designed to appeal to the better scholars. The program includes an internship in lieu of student teaching. The internship is a one-semester, salaried, licensed teaching position in a public school. Admission to the program requires acceptance by the graduate school, a bachelor's degree from an approved institution with a major in agriculture, and an interview by the agricultural education department. The student may choose to seek certification only, rather than complete the entire program for the master's degree.

Special Post-Master's Degree in Education. Of growing importance in education is the professional degree beyond the master's for teachers who do not wish to seek the doctorate. Fifteen institutions offer such a program in agricultural education. There have been some problems in preventing this type of degree from gaining a reputation as a resting place for persons who tried for the doctorate and failed, but the problems of this kind seem to have been solved, and the degree has achieved a respectability and status of its own.

The growth in popularity of the special post-master's professional degree has been quite spectacular, especially since other special degrees appear to be losing favor. Eells reports that "unlike the various special types of degrees which appear to be disappearing, a new degree in the professional field of education has developed since 1950 and appears to be growing more popular. It is intermediate between the master's and doctor's degrees, usually representing two years of work beyond the baccalaureate, and frequently designated as a 'sixth-year degree (12.14).'" The name of this degree is not yet firmly established, but most agricultural education programs refer to it as an "educational specialist" degree. Eells further explains that this specialist degree was first offered by the University of Kansas in 1950 and granted there in 1954, and that several other institutions offered sixth-year degrees under other titles as early as 1933. At the University of Massachusetts it is called "Certificate of Advanced Graduate Study," and is designed "to provide an intensive, cohesive program of professional development for education specialists beyond the master's of education degree."

Two years of work beyond the bachelor's or one year of work beyond the master's is required for the granting of the various specialist degrees. Programs are planned with an advisor or a committee to meet the needs of the individual student.

In some institutions, there are no specific course or other requirements to be met beyond the total hours-of-credit requirement. One institution requires some research emphasis and a comprehensive oral examination. In yet another institution, approximately one-fourth of the program must consist of field work, internship, externship, and similar laboratory experiences. Some institutions permit transfer of a limited number of hours of credit from other approved schools.

Doctoral Level

There are 18 institutions which reported offering doctoral-level work with a focus in agricultural education. These institutions reported 167 enrollees (more than 50 percent of whom are part-time), and during the 1978 school year, they awarded 46 Ph.D's in agricultural education. Approximately 10 of these individuals who earned Ph.D's entered new job streams. The others were likely on leave from jobs, or they were international students who returned to their home countries. While several disciplines, such as history and literature, report a surplus of persons with new doctoral completions, this apparently is not true for agricultural education. Employment opportunities are still plentiful for those who complete the doctoral degree.

Several options are available for the student with an agricultural education background who wishes to earn the Ph.D. or the Ed.D. Some institutions offer a pure agricultural education program. In such a program, the core courses are in agricultural education. Others report an agricultural education/vocational education core. As implied, there would be additional study in some advanced agricultural education courses and in more broadly based courses in vocational education. Still others report a doctorate in vocational education. Here, the entire study program is likely to be the more widely defined courses. Also more than one combination is possible at several institutions. This diversity is regarded as a strength of the profession, but it does necessitate that potential doctoral students know the type of program desired to achieve their professional goals.

The doctorate is sought by an ever-increasing number of students. Those who can qualify for it and those who have earned it are in demand not only in colleges and universities but also in industry and government. A significant number find employment in community colleges, technical schools, and even secondary schools. Additionally, there are a small but important number of persons holding the doctorate who teach vocational agriculture at the secondary school level.

Selecting the Program and Degree

As pointed out earlier, the type of master's degree or doctoral degree is not an important distinction for most students. In some foreign countries, however, a Master of Science might be acceptable, while a Master of Education would have little value because of tradition. There are, nevertheless, factors the teacher should consider in selecting the degree and program to seek. Some of these are the following:

1. Personal preference and goals. Some individuals may have a definite preference for a particular degree. Providing there are no other factors of more importance which dictate the choice of degree, the individual should by all means choose the more attractive program. The student seeking an advanced degree should have a reasonably clear idea as to why he/she desires the degree. That is, the person should know his/her personal goals. Not all institutions have a program, for example, enabling one to specialize in agricultural mechanization. Further, Carter has identified 14 competencies that are critical for a teacher educator (12.15). Thus, if one wishes to acquire a doctoral degree as a teacher educator, it would be wise to consult an institution's list to determine the extent to which these competencies may be acquired in the programs being considered.

2. Location of programs. Some graduate programs of agricultural education are located in colleges of agriculture, and others are located in colleges of schools of education. The location will clearly influence the type of programs available at the institution and may influence the degree as well. The November 1977 issue of The Journal of the American Association of Teacher Educators in Agriculture provides a perspective of this, debated by teacher educators Binkley (12.16) and Knebel (12.17).

3. Emphasis on research. As indicated in the descriptions of the various kinds of degrees offered, some degrees permit or require a greater involvement of the candidate in research through courses in statistics and research. Some degrees demand or permit more emphasis by the candidate on problems courses and independent study than do other degrees. Teachers who wish to become involved in doing research as compared with becoming able consumers only should consider carefully what the degree selected provides regarding research requirements or possibilities.

4. Thesis requirement. Many master's degrees contain a thesis option, while others either require or do not require a thesis. Only 7 agricultural education departments reported that a master's thesis is required, while 40 indicated the thesis is optional, and 14 reported that the master's degree is in coursework with no thesis possible. This is often related to the title of the degree. A Master of Science degree will, for example, require a thesis, while the Master of Art in Teaching is in coursework with no thesis. The candidate who wishes to write a master's thesis needs to be aware of the options at the school selected or being considered. At the doctoral level, the options are different. The Ph.D. requires a research dissertation, and the Ed.D. demands a major project which often is a research dissertation. Some institutions insist upon an intensive field project.

5. Residence requirements. The residence requirements vary for the different kinds of degrees. If a teacher wishes to earn part of the credits needed for the degree through enrollment in off-campus or extramural courses, he/she obviously should not seek a degree with the requirement that all work be done as a full-time student on campus. Nearly one-third of the institutions reported no residence requirement for master's study in agricultural education. The longest was one year of study at the institution. Nearly all institutions permit some of the coursework and residence work to be

transferred into the student's graduate program. The amount of credit which can be transferred is negotiable, while the amount that must be earned at the degree-granting institution is specified.

Doctoral programs more typically require six semesters of graduate work beyond the bachelor's with two of them being consecutive. This gives one year in residence as a full-time student. Also, usually a time space of one year must pass after the preliminary examinations are completed before the degree may be granted. This time is usually spent on research or field projects.

6. Flexibility of program planning. Although the choice of the major area of study gives direction to much of the course planning involved in a degree program, some degrees permit a wider range of selection in courses than do other degrees. The teacher who wishes to have maximum freedom in the selection of coursework should seek a degree which permits that freedom.

7. Relationship to further graduate work. Although most master's degrees are automatically accepted as a step toward the doctorate, the same cannot be said of the degrees which are between the master's and the doctorate. The work for the intermediate specialist degree may simply not be accepted as part of the credit needed to earn the doctorate, regardless of the quality of the work done. Teachers who have any idea that they may eventually seek the doctorate should consider this factor in planning their graduate programs. Again, the student needs to be aware of his/her professional goals. A person wishing eventually to get a doctorate in a subject matter field such as animal science or agronomy should find out if the Master of Arts in Teaching degree is acceptable. This note is added because sometimes this degree may be the only one available from an agricultural education department.

8. Assignment of advisor. Often a teacher will wish to do his/her graduate work under the direction of a particular member of the faculty. The teacher should make certain that the faculty member with whom he/she wishes to work is approved for advising candidates for the degree chosen.

Education vs. Agriculture as the Graduate Major

One of the major decisions to be made by teachers planning graduate programs at the master's level is that of selecting the major area of study for the degree. For teachers of agriculture, the problem is usually one of choosing between a technical agriculture major and a professional education (agricultural education) major. Either major could be appropriate because of the nature of the teacher's job. Further, the decision regarding the choice of the major area of study often also results automatically in a decision regarding the kind of degree the teacher will seek. Since this is one of the more critical decisions the beginning graduate student must make, the advisor has a responsibility for helping the student consider carefully the several factors which may affect the decision. Some of these factors suggest that a major in education would be best; others indicate that a technical agriculture major would be most appropriate. The following factors should be considered in selecting a major.

1. Need for professional competence. Perhaps the primary factor in the selection of the graduate major is the need of the teacher for professional competence. Regardless of the long-range plans of the teacher, it is critical that he/she do well as a teacher for whatever period of time he/she may serve in that capacity. This means that priority must be given to maintaining and enhancing professional competency. This can best be done through organized instruction in professional education courses in agricultural education and in areas such as educational psychology and educational philosophy. For the teacher of agriculture, the professional education major should be in agricultural education. The teacher who wishes to begin preparing for non-teaching job opportunities in agriculture as a part of his/her degree program would best be advised to use a minor for that purpose.

2. Need for technical agriculture competence. Although professional competence--the ability to teach--is the most critical factor in success in teaching, a knowledge of the subject taught is also critical. However, the needs of a teacher of agriculture for competence in technical agriculture may be better met through a degree program with an agricultural education major than through a degree program with a technical agriculture major. The technical agriculture major often requires a concentration of effort in one field of agriculture almost to the exclusion of other technical agriculture areas. Although teaching areas in agriculture are becoming more specialized, the teacher of agriculture must maintain competence in several areas of technical agriculture. Most degree programs with a major in agricultural education provide for the needs of the teacher of agriculture in technical agriculture.

3. Need for background or prerequisite courses. The major interest of the person preparing to teach is in agricultural education. As is true for other undergraduate majors, the requirements to be met place some limits on the amount of coursework which may be taken as electives. The person who later decides to change his/her major when beginning graduate study may find that some undergraduate work must be taken without credit before he/she can go very far in his/her new field.

4. Ability to adjust to changing programs of vocational agriculture. Programs of vocational education in agriculture are changing in direction. One of these directions is the creation of multiple-teacher departments, with teachers of agriculture specializing in various areas of technical agriculture. This specialization may be accomplished through minors and elective courses. Another change is the rapid development of post-secondary programs in technical agriculture, which provide opportunities to teach in one or two technical agriculture areas.

As long as teachers of agriculture remain in teaching, they should continue their studies as graduate students, with majors in the areas most closely related to their work--agricultural education. They can strengthen their teaching competencies in technical agriculture areas by carefully selecting minors and elective courses.

Research in the Graduate Program

Emphasis on research in graduate programs in agricultural education has been stimulated by the escalation of federal support for education. This increased emphasis has been felt especially at the master's level, due to the employment of master's degree candidates as research assistants. However, there continues to be concern, which has existed for many years, regarding the research orientation of teachers. Simpson states:

> An over-all review of hundreds of research studies reveals that almost invariably teaching and learning test out better if teachers are actively researching themselves. The particular nature of the investigation into methods or approaches that the teacher carried on does not seem to be of paramount importance. The key factor is this: The teacher is consciously and carefully studying and investigating his procedures and trying to improve them (12.18).

Research Competencies Needed. Perhaps the most important implication in regard to research in graduate programs is the need to identify the research competencies the teacher needs. According to Simpson, the nature of the investigation conducted by the teacher is not of paramount importance (12.19). This implies that the research competencies of significance lie in the realm of understanding the value of research in teaching. If teachers understand how the results of research can help them to improve their performance, then they will conduct research.

If teachers are to conduct research, they need to know how to identify and define problems to be researched and how to formulate hypotheses to be tested. Further, they need to know how to design or structure the research and how to collect, summarize, analyze, and interpret data. The techniques of tabular analysis and the more easily calculated statistical tests of significance are useful to the teacher-researcher. Developing the research competencies indicated would also contribute to developing the ability to interpret and use the research done by others.

Instruction About Research. Instruction designed to develop certain research competencies should be provided in most graduate programs. The emphasis needs to be on teaching research tools and techniques which a teacher who is conducting a full teaching program can utilize, rather than on teaching those tools and techniques which will be used by the paid researcher who has modern computers to aid him/her. For most graduate programs, this will require a reorientation of program objectives to a point somewhere between the development of the highly specialized researcher and the ignoring of the need for teacher research competency. Thus, it may be necessary to design new courses, since the present courses may still be needed for preparing the research specialist.

In addition to providing special courses on research, graduate programs should allow for the consideration of problems needing research, as well as the consideration of the use of research which has been conducted, in their graduate courses.

Conducting Research. Graduate programs should also include teacher-conducted research as a planned part of the total program.

Although the thesis option is, on the surface, designed to provide experience in research, teachers do not have to use the thesis option to obtain a sound research experience. Many teachers simply do not have any desire to prepare a thesis and would resist efforts to advise them in that direction. For such teachers, other forms of research experience should be provided. These forms include problems, field studies, or independent study courses. The design, conduct, and reporting of the research of teachers in these non-thesis research courses can be as rich and meaningful a learning experience as that provided by the writing of a thesis.

The primary factor in the conduct of research by the teacher-graduate student is that the research be on the problems of the teacher and that it be the kind of research that will be expected when the student is employed as a full-time teacher.

Reporting Research. As teachers conduct research, it is important that ways and means for publishing the results of their research be identified. For this purpose, both journal articles and departmental publications should be fully utilized. Arrangements for publishing the research report should be made a part of the research plan.

Development of the Teacher of Agriculture as a Scholar

Hamlin expresses a general concern in the following:

> We have already tried to go further than we can go in making teachers of agriculture purveyors of small bits of technical information and teachers of small skills to an ever-growing constituency. The end of the road we have been traveling is a dead end. We shall, sooner or later, have to retrace our steps. When we find eventually that we must reconceive the task of a teacher of agriculture, we should give a high priority to his functions as thinker and scholar. . . . We need to be alarmed when we hear teachers of agriculture saying that they no longer have time to read, or think, or share the conversations of the thoughtful people in their communities (12.20).

Indeed, when teachers eagerly seek out prepared teaching plans, when they suggest that state and even national courses of study be developed, rather than create their own, this is cause for alarm. In many ways, teachers of agriculture are becoming highly skilled educational technicians rather than creative professionals. A new regard for the art and science of teaching must be aroused. It is a major responsibility of the graduate program to make certain that the polishing of the technical skills aspects of teaching does not become the major goal of the teacher as a graduate student.

The place of the conduct and reporting of research in the graduate program has already been discussed. Improvement in this aspect of graduate programs can contribute much to the development of a keen and analytical mind. It can create an awareness of and an ability to recognize the major critical problems in agricultural education and the means available for investigating them. A teacher-oriented research emphasis would do much to develop the teacher as a scholar.

Second, graduate programs need to have provision so that students can present and debate their philosophy of education with other students and faculty members. The ability to expess one's thoughts clearly and concisely, the ability to defend one's ideas, and the ability to analyze critically the ideas of others are not developed in intellectual isolation. The challenge of debate with an intellect the graduate student respects is required. Through this kind of debate are dreams refined into workable and testable hypotheses.

Third, graduate programs must allow graduate students time to read critically the writings of the leaders and philosophers in their major fields, as well as in related fields. It is this kind of reading that leads to the germination of the ideas that become dreams, dreams to be tested again and again in debate with other students and teachers. It is this kind of reading that transports graduate students from the fields in which they work to the battlefields on the perimeter of their profession.

Finally, graduate students must be encouraged, even required, to express their thoughts and ideas in writing. Ideas expressed in oral debate can be forgotten, modified, or repudiated. The idea placed on the printed page must stand or fall according to its merits. It cannot be rescued or denied once it has been placed for others to see.

The development of the teacher of agriculture as a scholar is the most important task, and the most neglected task, of graduate programs in agricultural education. The future of agricultural education at all levels is more dependent upon the development of creative thinkers in agricultural education than upon any other factor.

Planning for Graduate Study

Continuing education has always been part of the life work of those in the teaching profession. It may be easier for teachers to accept this "fact of life" now that most other persons also are faced with the need for continuing their education beyond the bachelor's degree. It remains only for them to plan the details of their education programs. Since one of the major problems in planning for graduate study, determining the degree for which to work, has already been considered, the other decisions which must be made remain to be considered. Serious students, particularly those seeking the doctorate will find help in consulting the latest edition of reference books such as <u>The Random House Guide to Graduate Study</u> and <u>Graduate Education: A Critique and a Program</u>.

Selecting the Institution

One of the trends in the last 10 to 20 years has been the rapid expansion of graduate programs in agricultural education. Many of the new programs are in small schools of agriculture. This is further evidence that the perspective student needs to be fully aware of why he/she desires graduate study.

Institutions, even those in the same state, differ widely in subjects available, admission standards, and degree requirements.

When to Start

The non-degree programs are considered elsewhere in this publication. For the degree candidate, planning must begin early in the college years. The institutional requirements for admission to graduate school make it imperative that students be aware of what they must do "to keep the graduate school door open" until they are ready either to walk through it or to close it. Students who are aware that their undergraduate records will be the major factor in determining eligibility for graduate study may be motivited to work just a little harder to achieve the necessary quality of performance.

Either as a full-time teacher who is taking off-campus courses or as a full-time student, an individual should begin taking graduate courses during the first year after he/she has earned the bachelor's degree. There are advantages and disadvantages to each approach.

The Teacher-Student. For the student who decides to take a teaching position, off-campus courses in agricultural education are especially beneficial during the first year. Several institutions offer special courses for first-year teachers to provide them with the help and guidance of the university staff and the benefits of discussions on common problems with other first-year teachers. The graduate student will probably never again have as many "teachable moments" or as much motivation for learning as he/she has during the first year of teaching.

Some other advantages of accepting a teaching position and taking graduate courses part time are as follows:

1. When individuals who have teaching backgrounds engage in graduate work, it is more meaningful to them. They can concentrate on actual teaching problems in class discussions.

2. As individuals teach, they become aware of needs and gaps which existed in their undergraduate programs. This knowledge aids them in planning their graduate programs.

3. As graduate students engage in research activities, their teaching experiences provide invaluable knowledge of problems faced by others in the profession.

4. Many individuals have a great satisfaction in becoming working, contributing members of society after so many years of study.

5. Many individuals have to consider providing for their families.

6. Some individuals feel that particularly desirable teaching positions which are available at the time of their graduation might not be open the next year.

7. Some individuals may need a year of teaching to make certain they wish to make teaching a lifetime career. This experience, if needed,

should precede their beginning full-time graduate study, especially if they anticipate a change in occupation.

The major disadvantage of working for an advanced degree while being employed as a teacher is obvious--it is a long process, and it consumes the time the teacher would otherwise be able to devote to his/her family, home, and recreation. Another disadvantage is that the off-campus student does not have access to university library facilities and other campus resources. This is partially compensated for by the usual requirement that one-half or more of the work for the advanced degree must be earned on campus.

The Full-Time Student. There are two kinds of full-time students to be considered: The student who begins study immediately after completing work for the bachelor's degree and the person who teaches for several years before starting graduate work on a full-time basis.

Those students who turn immediately to graduate work usually consider that:

1. It is less of a burden to themselves, and to their families, if they earn the advanced degree before taking a teaching position.

2. It may be difficult to tear up roots and return to campus after having settled in a community.

3. Assistantships or other forms of financial aid offered at the time of graduation might not be available when they are ready to return to school.

4. The additional year of study adds to maturity and judgment, thus enhancing the individuals' chances of success in teaching.

5. The additional year of study adds to an understanding of teaching and to a knowledge of subject matter, thus increasing the individuals' chances of success in teaching.

6. Full-time study on campus offers a richer experience because of the full availability of all the resources of the university.

7. Most school systems have a step-wise pay system enabling individuals with master's degrees to start teaching with higher salaries. Thus, the cost of obtaining a master's degree is quickly repaid.

The disadvantages of full-time graduate study prior to obtaining teaching experience are obviously that the individual students do not have the background of experience to give full meaning to their graduate studies, and they do not know their own needs, which can be gained from experience to use as a basis for planning their graduate programs. Some persons may also have to reject opportunities to secure employment in particularly desirable teaching situations, opportunities which may or may not be available the next year.

Those individuals who teach for several years and then return to campus to engage in graduate work as full-time students have the advantage of experience, of full access to the resources of the university, of the maturity of judgment developed over the years, and of the financial stability which comes from full employment. If the students have, in addition, kept alert by taking off-campus courses, then they will probably be in a position to profit from graduate studies to the maximum of their potential.

There is, of course, no one best time to start graduate work which will fit all persons. Each person must consider his/her own situation and arrive at the best decision possible for himself/herself.

Financial Aids

Financial aid for full-time graduate study in agricultural education may be available in several forms. The most important type of aid available is the graduate assistantship, usually designated as teaching assistantship or research assistantship. These assistantships pay a stipend, plus waiver of tuition and fees, for 15 to 20 hours per week. Since 1963, every year the January or February issue of The Agricultural Education Magazine has contained information about graduate assistantships available to agricultural education majors.

The graduate fellowships, another form of aid, are limited to students with exceptionally high grade-point averages. Students who have been granted certain kinds of fellowships perform no work, and, indeed, they are not permitted to work.

Some summer institutes also provide financial support for teachers and their families through federal grants designed to encourage teachers to engage in professional improvement activities. These "stipends" require no service or work by the recipients other than the studies connected with the institutes or workshops.

The student who accepts an appointment of any kind should be aware of the limitations placed on the amount of coursework which can be taken. Appointments for more than a quarter-time usually add to the length of time it will take the candidate to earn the degree.

Financial aid is discussed in more detail in Chapter 10.

Issues Relating to Graduate Study

Many issues regarding graduate education for teachers have yet to be solved. New issues emerge, but few of the old ones disappear. An issue is defined here as "a recognized problem on which two or more definite positions have been taken by significant segments of the profession." In a survey undertaken to get up-to-date information on graduate education, teacher educators were asked to identify one or two current issues of the profession. More than 35 distinct issues were identified. The 10 judged to be the most significant are discussed on the following pages.

Issue No. 1--Delivery System. How can teachers of agriculture acquire needed graduate courses? When graduate education was first introduced in 1847 at Yale with 11 students, it was assumed that all students would be full-time on campus. The same was true for post-graduate education, introduced by the opening of Johns Hopkins in 1876. No thought was given to the fully employed person who might need graduate courses, let alone need a graduate degree (12.21). It could not be determined in the survey when graduate education was altered to permit the fully employed person to pursue graduate studies. For graduate study in agricultural education, the major effort in terms of number of students is with the part-time student.

For doctoral students, the delivery system is not a major concern. Most students pursuing the doctorate probably begin their degree programs as full-time students. Some may later change to part-time status.

For master's students, however, only a small percentage leave their teaching positions to enroll as full-time graduate students. They expect to attend late-afternoon or evening courses and summer sessions. Summer sessions of six to eight weeks in length are no longer adequate. Secondary teachers find it increasingly difficult to be gone from their communities for such long periods of time. Universities have often found it necessary to shorten summer sessions and to offer more credit in workshop formats for one to two weeks in length.

One quite new delivery system tried in the fall of 1978 at The University of Wisconsin - Madison was a "weekend program." Students came to campus every third weekend. Classes were held Friday evening starting at 5:00 P.M., and on Saturday until 3:00 P.M. Eighteen credits were to be obtained over a two-year period. Forty-six students began the program the first semester, and the number increased to 71 the following semester. Of these, nine were teachers of agriculture at the secondary or post-secondary level.

An additional concern for delivery involves technical agriculture courses. Several persons who report this as an issue are concerned that it is increasingly difficult to get technical agriculture courses available during their summer offerings. The implication is that the courses are needed in their graduate programs if quality is to be maintained.

Issue No. 2--Flexibility, or the Lack of It. The first respondent who identified this as an issue said, "Required courses leave little room for flexibility." Several other comments that followed were of a similar nature. With one exception, this issue existed in smaller maturing, non-land-grant institutions that offer the master's as their highest academic degree. Apparently these institutions require specific courses or types of courses in technical agriculture education courses.

The exception was a university offering the Ph.D., which described the required courses as being in vocational education and supporting education foundations. Again, the opinion offered led to the same conclusion: too little room for agriculture education courses in the master's program.

However, it should be pointed out that the former example was for a master's degree in agricultural education, and the latter was for a master's degree in education.

Issue No. 3--Research at the Master's Level. In 1963, Eells wrote:

> Various problems regarding the true function and possible modifications in the requirements for the master's degree and the best uses for it have been under discussion at intervals during

the twentieth century. Should the master's degree be regarded as a terminal degree, significant in itself? If so, for what purposes or positions? Chiefly for secondary-school teachers, or for college teachers? Should it be concerned with subject matter or should it include pedagogical methodology? Should it be thought of merely as a stepping-stone on the way to the doctorate? Should the period of study for it be increased to two years? Should a thesis or dissertation be required? These and related questions have been the subject of debate for almost a century, and no general agreement has been reached on most of them (12.22).

Today, the question of master's thesis research is still unresolved. Those who support the thesis see research as being necessary to the profession. Each professional is viewed as either a generator of research or a consumer of research. The thesis is intended to help provide the needed competencies for either or both roles. It can also be used as a training ground for those who go on to pursue the Ph.D.

On the other hand, others propose that the master's degree should have an identity of its own and not be tied to other degrees (12.23). Because it serves a variety of functions for varying client groups, it may be appropriate for it to be a professional degree which helps individuals be better teachers (12.24).

While there is some controversy as to whether a thesis should be a required part of the master's degree program, the larger issue is determining what is the "purpose" of the master's degree program.

Issue No. 4--Technical Agriculture. The focus of this issue is not whether or not technical agriculture should be a part of master's level graduate work but the amount of it that should be provided. Small programs in schools of agriculture appear to desire far fewer technical agriculture courses than is now required. Others report they would like to require more technical agriculture courses but that such are not available to their students--particularly in summer course offerings.

Issue No. 5--Financial Assistance. This is more of a problem than an issue, but it has a significant impact on doctoral programs. Traditionally, most of the support for doctoral students has come from research monies for experiment stations, with the U.S. Department of Education being the primary source of such research funds. The primary purpose of these funds is research, not the support of advanced graduate students. These and other funds are remaining at the same level or decreasing at a time when some doctoral programs need to be expanded. Whether or not doctoral programs should be so closely tied to research funds is perhaps the real issue.

Issue No. 6--Focus on the Graduate Program. Although related to Issue No. 4, this is a broader issue since it also includes doctoral programs. A number of problems which surfaced are grouped here. These include: "How can we draw from other fields (psychology, sociology, philosophy) and still have an agricultural education core?" "How specialized should the Ph.D. be?" "Should the degree program be in agricultural education or vocational education?" "Should ours be a professional degree?"

Issue No. 7--Standards for Admission to Graduate School. The trend toward universities raising grade-point-average requirements for admission to graduate school, as noted by Krebs (12.25), seems to have abated. However, graduate schools now are permitting fewer exceptions to pre-set standards. The undergraduate grade-point average, the most universal admission criterion employed, as described earlier in this chapter, apparently began to be used at a time when persons moved directly from undergraduate programs to advanced studies in medicine, law, engineering, chemistry, genetics, etc. Undergraduate grades were the best predictor of how successful a person would be in graduate school.

However, the majority of agricultural education graduate students today do not go directly into graduate school. Some of them spend 3 to 11 years as practicing teachers before they begin graduate work. In such cases, the undergraduate grade-point average is not necessarily an accurate indicator of success in graduate school. Other important factors are experience and maturity. But, knowing how to apply these and other factors as admission criteria is lacking, as is knowing how to communicate these judgments to colleagues who make graduate school policy and who believe that the undergraduate grade-point average should remain the primary criterion for admission to graduate school.

The other issues identified in the survey were Issue No. 8, graduate credit for inservice work taken by teachers; Issue No. 9, professional experience before advanced degrees can be granted; and Issue No. 10, the upgrading of FFA and leadership skills of secondary teachers.

CHAPTER 13

Evaluation of Teacher Education Programs

Alfred J. Mannebach
UNIVERSITY OF CONNECTICUT

Charles C. Drawbaugh
RUTGERS UNIVERSITY

Evaluation, defined here as the determination of the worth of some phenomenon, is a vital part of any educational endeavor. Just as local programs of vocational agriculture are assessed, teacher education programs should also undergo evaluation. This conclusion is supported by the wealth of material in the literature regarding the evaluation of teacher education in agriculture programs.

Although teacher education in agriculture programs operate under a generally accepted philosophy, each program is characterized by different purposes, environments, constraints, and procedures. These various component parts and their interrelationships contribute to the complexity of the evaluation process. However, they also point to the need for a unified approach to evaluation. A well-conceptualized and well-organized evaluation effort results in recommendations supported by evidence which identifies program strengths and deficiencies, as well as providing guidelines for improvements.

Definitions of "Evaluation"

Many definitions of "evaluation" have been proposed by educators. However, there is common agreement that evaluation is the assessment of the worth or value of a thing; an appraisal of some kind; the systematic and objective determination of the merit or worth of something. Bender, et al., states that "evaluation is concerned with placing values on processes, procedures, outcomes and activities (13.1)." Cross gives this definition: "Evaluation is a process which determines the extent to which objectives have been met (13.2)."

Cross also advances five principles of evaluation based on the evaluation being an integral part of the teaching-learning process. These principles are:

1. Evaluation should be in terms of selectives objectives.
2. Evaluation should be comprehensive.
3. Evaluation is a cooperative process.
4. Evaluation is a continuous process.
5. Evaluation is concerned with valuing.

Purposes of Evaluation

There are several reasons why teacher education in agriculture programs should be evaluated. The most important reason is to provide a solid foundation for decision making and program planning and program improvement. Evaluation is also conducted to improve staff performance and to ensure that programs are accountable. Other valid reasons for evaluating teacher education in agriculture are:

1. To justify the expenditures invested in the program.

2. To provide an objective and valid description of the program.

3. To establish benchmarks for future comparisons.

4. To serve as a systematic review and to identify areas of strengths and weaknesses.

5. To serve as a public relations mechanism.

6. To involve people in the evaluation and provide them with information about the program.

7. To motivate faculty and staff members.

Evaluations are conducted to arrive at answers. Equally important is the ability to transport the answers from the professional domain into the political arena. Many prominent agricultural educators believe that it is largely through the political process that knowledge is translated into social policies and programs and that resources are mobilized to support services benefiting clientele. Regarding the use of evaluation results, it is possible that major future breakthroughs in agricultural education will depend on how well teacher educators in agriculture do their job in the political arena.

Evans and Terry provide further insight into evaluation philosophy. They maintain:

> As new teacher education programs develop in vocational education, it is essential that systematic evaluations be made of these programs. The purpose of the evaluations is not to "prove" that a particular program is good or best. This is virtually impossible in evaluation as it is usually done. Rather, the purpose of evaluation is to provide information for making decisions about the program by the staff participants, by potential participants of the program, by those funding the program and by persons who might be interested in starting a similar program. The evaluation should describe well what is happening in the program and what the outcomes of the program are.

The evaluation should also be designed so that it provides data and information that will permit systematic examination of the assumptions underlying the program (13.3).

Anderson, et al., proposes the following characteristics of evaluation (13.4).

1. The primary purpose of evaluating an education or training program is to provide information for decisions about the program.

2. Evaluation results should be useful for program-improvement decisions, not just for decisions about continuation or termination.

3. Evaluation information should be provided in time to be useful for such decisions.

4. Evaluation is a human judgmental process applied to the results of program examination.

5. Evaluation efforts should take into account the long- and short-term objectives of the program.

6. Just as it is important to consider the effects that a program was not necessarily designed to foster, so is it important to delineate the events other than the program that might have produced any effects that are discerned.

7. It is difficult to conceive of a useful model for evaluation of education/training programs that is not multivariate in nature.

8. The processes of obtaining information for evaluation should meet appropriate criteria of objectivity, reliability, validity, practicality, utility, and ethical responsibility.

The ASCD yearbook asserts (13.5):

The test of an evaluation system is simply this: Does it deliver the feedback that is needed, when it is needed, to the persons or groups who need it? If a system of evaluation is to meet this test, it must satisfy several basic criteria:

1. Evaluation must facilitate self-evaluation.

2. Evaluation must encompass every objective valued by the school.

3. Evaluation must facilitate learning and teaching.

4. Evaluation must provide records appropriate to the purposes for which records are essential.

5. Evaluation must provide continuing feedback into the larger questions of curriculum development and educational policy.

General Principles

Some general principles important in the evaluation of education programs are:

1. Evaluation should start with a statement of objectives.

2. Criteria for judging or evaluating must be given early in the process.

3. A variety of evidence should be assembled for the evaluation.

4. Many instruments, rather than one instrument, should be used.

5. Objectives should be established and measured by those who are responsible for, involved with, and affected by the program.

6. Evaluation should be a continuous procedure; specific outcomes should be measured periodically.

7. Both objective numbers and subjective judgments are useful in evaluation.

8. Evaluation is concerned with context, input, process, and product measures.

Steps in Evaluation

Tuckman proposes a brief outline of appropriate evaluation steps (13.6). They include:

1. Specifying the outcomes and their measurement.
2. Specifying and evaluating inputs and process.
3. Constructing a design.
4. Carrying out the evaluation.

Knuti states (13.7):

> Getting accurate information about the extent to which we are doing what we set out to do is an essential step in evaluation. Evaluation requires a number of different operations and the following seem to be essential:
>
> - Stating the objectives of a group or organizational program in specific terms so that evidence of the degree to which objectives are being achieved can be attained.
>
> - Securing evidence of the degree to which objectives are being achieved.

- Securing facts about what is being done to achieve the objectives.

- Developing ideas about what factors might be helping and hindering the achievement of objectives.

- Securing evidence for and against these ideas.

- On the basis of evidence obtained, revising ideas about what is helping and hindering the achievement of objectives.

- Developing and trying-out methods of remedying weaknesses in the program.

To be effective, evaluations must be well-planned, organized, and coordinated. Tuckman explains:

> As important as it is to communicate schedules accurately, it is even more important to communicate expectations. Since you cannot anticipate every problem that will arise and therefore plan for it or eliminate it by scheduling, you must make school personnel aware of your expectations--the general way that you want the evaluation to proceed (13.8).

> Some specific procedural suggestions advocated by Tuckman are:

- Lay out a detailed schedule in your evaluation plan. Prepare a set of tasks to be completed . . . and assign each task a specific date.

- Prepare a list of responsibilities for everyone who is to be involved in the evaluation.

- Prepare an observation plan. If classroom observations are to be made, lay out in advance a detailed plan for which teacher is to be observed when, and by whom.

- Conduct briefings for all involved persons. Conduct them as necessary as the evaluation proceeds.

- Keep tabs. Do not assume that your plans will be carried out or your schedule properly implemented. Provide reminders.

Coster provides some additional viewpoints regarding evaluation of vocational-technical education, which includes teacher education in agriculture.

> Throughout the years, the evaluation of vocational, technical, and practical arts education has been more qualitative than quantitative, more subjective than objective, more introspective and impressionistic than empirical, more ex post facto than apriori and directed more to the process than to the product (13.9).

Guiding Principles

In 1958, the Teacher Education Committee of the Agricultural Education Division of the American Vocational Association set forth guiding principles for institutions training teachers of vocational agriculture (13.10). The establishment of principles was for the purpose of contributing to the development of training objectives within, and between, institutions. The seven principles set forth for teacher-training institutions are presented in Chapter 3. These seven guiding principles continue to be valid today for teacher education in agriculture.

Program Objectives

Chapter 4 focused on the importance of instructional objectives for preparing teachers of vocational agriculture. Graduate academic degree programs can be evaluated in terms of the objectives established for those programs. For example, graduate academic degree program objectives can be a meaningful source of comprehensive examination questions. The examination measures to some degree what the student has learned in terms of the objectives set forth. An evaluation of this one measurement by the teacher educator is a means of estimating or predicting how well the student will perform as a teacher. The measurement and evaluation of a larger sampling of students by the same means provides the teacher educator with an indicator of the quality of the academic degree program in delivering the instruction and prescribing the learning activities necessary to master the objectives. Therefore, if the program objectives are keyed to a valid job description and/or certification requirements, and if the teaching/learning responsibilities have been attended to fully, the output of the program should be competent educators at the academic level and in the specialty areas for which they have been prepared.

Accreditation

Accreditation is a process whereby an organization or agency recognizes an educational institution or program of study as being in conformity with a set of standards previously agreed upon. The functions of accreditation are as follows (13.11):

1. Certifying that an institution has met established standards;
2. Assisting prospective students in identifying acceptable institutions;
3. Assisting institutions in determining the acceptability of transfer credits;

4. Helping to identify institutions and programs for the investment of public and private funds;

5. Protecting an institution against harmful internal and external pressures;

6. Creating goals of self-improvement of weaker programs and stimulating a general raising of standards among educational institutions;

7. Involving the faculty and staff comprehensively in institutional evaluating and planning;

8. Establishing criteria for professional certification, licensure, and for upgrading courses offering such preparation; and

9. Providing a basis for determining eligibility for federal assistance.

The accrediting procedure usually involves five basic steps (13.12).

1. Standards. The accrediting agency, in collaboration with educational institutions, establishes standards.

2. Self-study. The institution or program seeking accreditation prepares a self-evaluation study that measures its performance against the standards established by the accrediting agency.

3. On-site evaluation. A team selected by the accrediting agency visits the institution or program to determine firsthand if the applicant meets the established standards.

4. Publication. Upon being satisfied that the applicant meets its standards, the accrediting agency lists the institution or program in an official publication with other similarly accredited institutions or programs.

5. Re-evaluation. The accrediting agency periodically re-evaluates the institutions or programs that it lists to ascertain that continuation of the accredited status is warranted.

The National Council for Accreditation of Teacher Education (NCATE) standards for the accreditation of teacher education (13.13), effective January 1, 1979, are divided into two parts: (1) basic programs and (2) advanced programs. The standards in both parts apply to all institutional programs leading to degrees or certificates, regardless of the location and time at which instruction takes place. Each standard has a preamble which gives rationale for the standard, interprets its meaning, and defines terms. The standards are used for evaluating institutional programs, including agricultural education.

An NCATE evaluation consists of applying 24 standards to basic teacher education programs and 25 standards to advanced programs at institutions of higher education. At both levels, basic and advanced, the standards are in the areas of (1) governance, (2) curricula, (3) faculty, (4) students, (5) resources and facilities, and (6) evaluation, program review, and planning. The standards are written by professionals knowledgeable about higher education, evaluation, and accreditation. They are offered as "minimum standards" for accreditation acceptability. While the NCATE standards are written for all programs and are stated in broad terms to provide flexibility to accommodate programs and institutions, they serve as important and even critical guidelines for the evaluation of agricultural education across the nation.

Haberman and Stinnett, after making a thorough analysis of the NCATE standards being used through the 1970's, reached the following conclusions.

> The NCATE standards are quite clearly the most comprehensive guides ever developed for assessing educational training. They provide a sound basis for judging present programs, as well as systematic guidance for institutions engaging in self study and self-improvement.
>
> The glaring shortcoming of these standards--their frequent lack of specificity--may in the final analysis be their greatest strength. Driven by budgetary necessities, professional pressures, and public scrutiny, institutions may very well establish higher minimum standards and expectations than those set at a national level (13.14).

It would appear that the conclusions reached by Haberman and Stinnett in 1973 are just as appropriate for the standards implemented on January 1, 1979. It was their position that standards will be as effective and as high as teacher educators make them.

Certification

For purposes of this discussion on the evaluation of teacher education, accreditation and certification are two sides of the same coin. They are complementary processes whose functions overlap considerably. Both accreditation and certification establish and maintain standards for evaluating the teaching profession. In general, accreditation applies to the entire educational program, while certification makes use of accreditation for the credentialing of teachers and administrators.

The Standards for State Approval of Teacher Education, published by the National Association of State Directors of Teacher Education and Certification (NASDTEC), is designed to assist state education agencies to approve programs for the education of teachers. Graduates of approved programs are certified by the state. NASDTEC explains the process in this manner:

> The approved program approach to teacher education and certification in effect in most states involves (a) the development of programs of teacher education by an institution in accordance with the established standards; (b) the official review and evaluation of each of the proposed institution programs in terms of the established standards and procedures by the state education agency and the

subsequent approval of the programs if the standards are met; and (c) the understanding that the teacher candidate, upon successful completion of a program thus approved, as attested by the institution, will be entitled to official recognition by the state education agency (13.15).

All teacher education programs shall be visited for approval every five years, or more often if deemed necessary.

The NASDTEC standards for state approval of teacher education are categorized as follows: (1) organization and administration of teacher education, (2) curriculum principles and standards for basic programs, (3) curriculum principles and standards for advanced programs, and (4) innovative and experimental programs. The curriculum principles and standards for the basic program in agriculture, and possibly for approving competency-based or performance based programs, will be included here.

The NASDTEC standards pertaining to programs for preparing teachers of agriculture are (13.16):

Standard I. The program shall provide that practical farm or other agricultural experience is a part of the requirements to be met for completion of the program leading to certification.

Standard II. The program shall provide an understanding of the biological, physical and applied sciences as they relate to practical solutions of agricultural problems.

Standard III. The program shall provide study of the essentials for production agriculture and the breadth in technical/agricultural industry.

Standard IV. The program shall include a sequence of studies and experiences which provide basic knowledge in areas such as the following:

A. Plant science and technology
B. Animal science and technology
C. Agricultural business management and technology
D. Agricultural mechanics science and technology

Standard V. The program shall provide for specialized preparation in one or more of the following occupational areas:

A. Agricultural production and marketing
B. Agricultural equipment and supplies
C. Agricultural products
D. Ornamental horticulture
E. Agricultural resources
F. Natural resources management
G. Environmental development
H. Forestry

Standard VI. The program shall provide leadership development including skills necessary in the development of agricultural youth organizations as a means of teaching leadership skills through study and practice of speech, parliamentary procedure, and group cooperative effort.

Standard VII. The program shall provide studies and experiences which enable the prospective teacher to perform the appropriate occupational skills while working with pupils and adults in projects and programs relative to agricultural instructional areas.

Standards Developed by Agricultural Educators

Major national publications, activities, and events which have fostered guidelines and standards for the evaluation and improvement of teacher education in agriculture in recent years include:

1. Guidelines for Developing Programs in Agricultural Education for the 1970's (13.17). These guidelines were generated at the May 1968 National Outlook Conference in Agricultural Education held in St. Louis, Missouri.

2. Transitions in Agricultural Education Focusing on Agribusiness and Natural Resources Occupations (13.18). These findings were the substance of a report of a national seminar conducted in May 1971 at Denver, Colorado.

3. Teacher Education in Agriculture Guidelines (13.19). This 1976 publication was developed by a committee of the American Association of Teacher Educators in Agriculture.

4. Standards for Quality Programs in Agricultural/Agribusiness Education (13.20). These standards were initiated in a national seminar held in Kansas City, Missouri, during the spring of 1976, and they were disseminated for use by the profession in 1977. They provide a basis for program evaluation.

Guidelines for Developing Programs in Agricultural Education for the 1970's. In this report there are 22 guidelines in teacher education which are generated under 5 philosophical concepts. These concepts are as follows (13.21):

1. There should be a common core of professional education content at undergraduate and graduate levels in agricultural teacher education.

2. Requirements for graduation and certification of prospective teachers in agriculture must be flexible.

3. Inservice programs should be based on essential professional and technical competencies.

4. Special attention should be placed on recruiting prospective teachers of agriculture.

5. Specialized national and regional programs must be formed to enable teacher educators in counsel with state departments of education officials to develop guidelines and give direction to teacher education in agriculture.

Transitions in Agricultural Education Focusing on Agribusiness and Natural Resources Occupations. In this publication the section on preparing and improving professional personnel deals with both preservice and inservice training. It does not treat all the functions of teacher education in agriculture but rather emphasizes (1) the recruitment of trainees, (2) the preparation of advisors of student organizations for leadership and personnel development, and (3) the commitment to research (13.22). Action steps are suggested for attaining each of the functions or objectives.

Teacher Education in Agriculture Guidelines. The AATEA criteria were developed for the purpose of evaluating and improving teacher education in agriculture programs. They were intended for consideration by NCATE in the revised edition of Standards for the Accreditation of Teacher Education. Professional organizations are given the opportunity to offer input for rewriting NCATE standards.

Eighty-one guidelines for evaluating teacher education in agriculture programs are categorized into the areas of (1) program of instruction, (2) faculty, (3) students, (4) resources and facilities, and (5) evaluation program review and planning (13.23). Additional details are provided in Chapter 3.

Standards for Quality Programs in Agricultural/Agribusiness Education. These standards were developed and validated by teachers, teacher educators, and administrators in agricultural education to serve as a model in the evaluation of programs and activities. The profession envisioned this project as the beginning of a long-range effort to upgrade vocational programs in agricultural/agribusiness education.

The standards are divided into 11 categories, including one which is directed specifically to teacher education. Teacher education, in turn, is further broken down into (1) recruitment, (2) enrollment, (3) counseling, (4) instructional program, (5) inservice education, (6) certification, (7) placement, (8) staffing, (9) support staff, (10) professional development, (11) research, (12) professional leadership service, (13) administrative organization, (14) financing, (15) facilities and equipment, and (16) evaluation (13.24).

Ninety-seven criteria are generated for evaluating teacher education in agriculture. Each guideline is checked in terms of exceeding, meeting, or not meeting an established standards in an evaluation. The judgment is then supported by written observations and followed with one or more recommendations.

Standards specific to quality teacher education in agriculture programs are as follows (13.25):

Recruitment

1. One agricultural education faculty member is responsible for coordinating the recruitment efforts of the department and

serves as a liaison with admissions personnel and others involved in the college recruitment program.

2. Emphasis is placed on the recruitment of students for agricultural education from various segments of the educational system including secondary and postsecondary agriculture/agribusiness programs, other departments of the college, and other colleges.

3. Off-campus people and organizations are actively involved in the recruitment of potential agricultural students.

4. Programs are conducted to promote and secure scholarships, fellowships, and financial assistance specifically to support agricultural education students.

Enrollment

5. Students enrolling in the agricultural education program are informed of the occupational experience and/or the supervised internship requirements as prerequisites to certification for teaching.

6. Means are provided to facilitate dual major students enrolled in an area of technical agriculture to simultaneously complete requirements for teacher certification.

7. The agricultural education faculty, in cooperation with the state staff, conducts annual needs assessments to determine new and replacement teacher requirements at both the secondary and postsecondary levels.

Counseling

8. At least ten percent of each full-time equivalent (FTE) faculty member's load is allotted to counseling and/or advising students.

9. The maximum number of advisees per FTE faculty is 25.

10. Undergraduates enrolled in other curricula who are considering the agricultural education program will also have an advisor assigned from the agricultural education faculty.

11. Students enrolled in a dual major, with one program being agricultural education, will be advised by faculty from both programs.

12. Students who plan to major in agricultural education are identified as early as possible for advisement, counseling, and placement purposes.

Instructional Program

13. The agricultural education curriculum provides for adequate preparation for teachers of vocational agriculture/agribusiness with course work distributed as follows:

General Education	20-30%
Professional Education	20-30%
Technical Agriculture/Agribusiness	30-40%
Electives	10%

14. The program of general education contributes toward making the prospective teacher mobile and adaptable with interests, appreciations, and understandings needed to participate effectively in agriculture and education.

15. The social, psychological, and educational foundations are required as an integral component of the instructional program.

16. Field experiences and methods courses in agricultural education are closely integrated in the professional preparation programs of teachers of agriculture/agribusiness.

17. Students enrolled in the teacher education program develop the following teaching skills:

    a. Using the problem solving method.

    b. Planning and conducting demonstrations.

    c. Planning and conducting field trips.

    d. Using community resources.

    e. Providing individual instruction.

    f. Using teaching aids effectively.

    g. Guiding students in selecting, planning, and conducting occupational experience programs.

    h. Supervising occupational experience programs.

    i. Organizing and using the FFA to strengthen the instructional program.

    j. Applying learning principles in teaching.

    k. Organizing and teaching postsecondary students and adults.

    l. Organizing and advising FFA, YFA, and other leadership developing organizations.

    m. Organizing programs for educating the handicapped and other special need groups.

    n. Providing career awareness and exploration programs in elementary and middle school grades.

18. Students enrolled in the teacher education program acquire the following management skills:

a. Organizing and using a local advisory committee.

b. Determining community and student needs.

c. Developing annual and long-range program plans.

d. Developing a course of study for the local department.

e. Arranging for adequate occupational experience programs.

f. Determining and acquiring needed instructional materials.

g. Determining and acquiring needed facilities and equipment.

h. Developing and using a filing system.

i. Organizing and operating a multiple-teacher department.

j. Evaluating students and programs.

k. Completing and analyzing annual reports.

l. Articulating secondary and postsecondary programs.

19. Cooperating schools at the secondary and postsecondary levels to be used for student teaching and other field experiences are selected by teacher educators in cooperation with the staff of the state agriculture/agribusiness education section (state staff).

20. Criteria to be used in the selection of cooperating schools is developed by teacher educators, state supervisors, and a select group of secondary and postsecondary teachers.

21. Cooperating teachers are prepared for their responsibility during an annual cooperating teacher conference sponsored jointly by teacher educators and the state staff.

22. A minimum of ten weeks of student teaching is required in the area for which certification is to be granted.

23. Observational experiences and/or other experiences are provided throughout the undergraduate program prior to student teaching and are supervised by the agricultural education staff.

24. Internships in supervised occupational experience for various periods of time are provided in specialty areas in agriculture/agribusiness in cooperation with selected employers to develop needed technical skills in teaching.

25. The teacher education staff provides leadership to undergraduates through the collegiate professional student organization by emphasizing the role of the teacher in professional organizations, to the administration and fellow staff members.

26. Each student teacher is observed and supervised a minimum of three times in the cooperating school by a teacher educator.

Graduate M.S. Level

27. Graduate programs provide for in-service training, professional development, and attainment of advanced degrees.

28. A flexible master degree program in agricultural education with and without a thesis is available.

29. Technical and professional graduate course offerings are available summers, during the academic year, and off-campus.

30. Graduate committees are chaired by agricultural education faculty members.

31. Graduate courses are designed and offered to meet special needs students.

Graduate Ph.D. Level

In institutions where resources are available and need exists.

32. Agricultural education, in cooperation with the graduate college, provides a doctoral program to prepare vocational teachers of agriculture/agribusiness education and other agricultural education personnel for leadership positions in such areas as teacher education, supervision, administration, research, and curriculum development.

33. Degree candidates are carefully screened.

34. Doctoral programs are flexible, permitting candidates to plan programs to meet individual needs.

35. Appropriate course work in applied research techniques is included in the academic program.

In-service Education

36. An in-service program for teacher improvement and the development of programs of vocational agriculture/agribusiness is provided.

37. Teacher needs are identified cooperatively by teachers, teacher educators, and the state staff.

38. Programs include both credit and non-credit offerings, on and off campus, throughout the state, and are planned on a long-range basis.

39. A special summer session of technical and professional courses, workshops, and seminars is conducted.

40. Faculty members in agricultural education participate with teachers and the state staff in planning and conducting an

annual state teachers conference, judging contests, state FFA convention, YFA convention, and other appropriate meetings.

41. Teacher educators provide individual supervision and/or consultive services as requested by the state staff.

42. An instructional materials service identifies, prepares, and distributes curriculum materials needed by teachers.

43. There are policies and means provided that enable selected teachers to assist other teachers with technical and professional matters.

44. A special credit course with at least 3 on-site consultation visits is provided for first year and returning teachers.

45. Credit and non-credit seminars and workshops are provided for postsecondary teachers.

46. Teachers of agriculture are certified only in the agricultural subject areas in which they were prepared to teach and have demonstrated competence.

47. Certification requires successful completion of an approved degree program in agricultural education and documentation of a prescribed period of approved occupational experience.

Placement

48. Seventy-five percent of those students completing the agricultural education program with a secondary or postsecondary teaching objective are employed as vocational agriculture/agribusiness teachers.

49. All agricultural education graduates become employed in the profession or in occupations in agriculture.

50. One agricultural education faculty member is assigned the responsibility for coordinating placement activities of graduates.

51. All agricultural education graduates are provided job placement counseling to assist them in becoming successfully employed.

52. Seventy-five percent of the students who initially enroll in agricultural education complete the requirements of that curriculum.

Staffing

53. Seventy-five percent of the agricultural education faculty have an earned doctorate degree; 100 percent have earned masters degrees in agricultural education, or the equivalent.

54. All faculty meet requirements for certification to teach vocational agriculture/agribusiness, including at least three years of successful teaching experience in vocational agriculture/agribusiness in the area or areas in which the faculty member is providing leadership.

55. Members of the agricultural education faculty have twelve-month appointments.

56. Faculty members have shown evidence of achievement in research and writing as measured by publications and research projects.

57. Faculty members have demonstrated leadership roles and are participating in professional organizations and state and national professional improvement meetings.

58. A minimum of two FTE faculty are employed to help students learn needed competencies in agricultural education, to advise students, and to supervise intern experiences. One FTE faculty member is provided for each ten degree/certification recipients (B.S., M.S., and Ph.D.). An equal number of FTE faculty members provide research and/or in-service functions.

59. A minimum of four (4) FTE faculty are employed to meet the technical education requirements of students in each of the following areas:

    a. Agricultural engineering and mechanics
    b. Plant and soil science
    c. Animal science
    d. Agricultural economics and business management

60. Comparable FTE faculty are assigned in specialized areas of certification (e.g.: ornamental horticulture, agricultural products and processing, forestry, and natural resources).

<u>Support Staff</u>

61. One full-time secretary is provided for each three FTE faculty members.

62. The teacher education faculty includes one graduate assistant for each FTE graduate level faculty member.

<u>Professional Development</u>

63. Each teacher educator participates annually in one or more recognized professional development meetings at the institutional, state, regional, or national level.

64. Each eligible teacher educator participates in sabbatical leave opportunities.

65. Each teacher educator makes one or more contributions annually to the profession in each of the following areas:

a. Research
b. Consulting
c. Scholarly writing

66. Professional development incentives are provided through promotion and tenure standards, merit raises, and awards for excellence in teaching, service, and research.

67. The teacher education faculty cooperatively develops and participates in an annual professional development program designed to strengthen teacher education.

68. Each teacher educator uses evaluation tools such as the _Standards for Quality Programs in Agricultural/Agribusiness Education_ and student and peer evaluations to determine areas in which professional improvement should be made.

Research

69. A minimum of 10 percent of total staff time is allocated to research activities.

70. A research advisory committee composed of representatives from agricultural industry, state staff, and teachers of secondary, postsecondary, and adult programs meet semi-annually to identify priorities and plan research efforts.

71. A planned system for dissemination of research results is used within the state, and existing systems (e.g.: ERIC, _The AATEA Journal_, _The Agricultural Education Magazine_, _Summary of Studies_, etc.) are used for national dissemination.

72. A research component is included in each graduate student's program.

73. The agricultural education faculty is active in securing outside funding for research efforts, including involvement with the state agricultural experiment station.

Professional Leadership Service

74. Service on university, college, and departmental committees is shared by all faculty members.

75. Departmental decision-making efforts rely on the use of committee work with evident results.

76. Staff members maintain active membership in professional organizations.

77. The department is active in agricultural education projects affecting program efforts at state, regional, and national levels.

78. Regularly scheduled joint meetings are held between teacher education staff(s) and the state staff.

79. Concurrence is obtained from the state staff prior to employment of new staff members in teacher education.

80. An annual written agreement with the appropriate state agency is in operation, indicating services to be provided by teacher educators and state staff.

81. Leadership is provided to organizations such as the Collegiate FFA, Alpha Tau Alpha, and Agricultural Education Club for students enrolled in teacher education in agriculture.

82. The teacher education staff provides service to the university faculty and other groups concerned with the improvement of instruction.

83. A vocational agriculture/agribusiness instructional resource center is maintained with resources and consultation services available to university personnel, secondary, and postsecondary teachers.

Administrative Organization

84. Separate identifiable administrative units within the university are responsible for teacher education in agriculture/agribusiness including pre-service and in-service teacher education, research, and service.

85. A department head is named and is responsible for the administration and supervision of the agricultural education program within appropriate university administrative structure.

86. An advisory council comprised of representatives of the state staff, secondary and postsecondary teachers, professional organizations, and related businesses and industries assists in formulating annual and five-year plans for teacher education in agriculture.

Financing

87. The university is responsible for providing adequate funding for staffing and staff development, program development, facilities, and resources to conduct a "high quality" teacher education program.

88. Major portions of funding are provided by the university with supplemental funds obtained from state and federal agencies and other sources.

89. Grants are obtained when available from U.S.O.E. and other sources for special innovative developmental curriculum and research projects.

Facilities and Equipment

90. Modern livestock, greenhouse, agricultural mechanics, and experimental farm facilities are used in the teacher education program.

91. The teacher education faculty in agriculture is provided adequate office and conference space and equipment.

92. An exemplary classroom of the type recommended for local agriculture/agribusiness programs is available for instructional activities in agricultural education. The classroom is adequately equipped with appropriate furniture, audio-visual equipment, and reference materials.

93. The institution provides a resource center which is used by the staff and students in agricultural education for the development of lesson plans and instructional aids.

94. Standard and contemporary reference materials in agricultural education, vocational education, and technical agriculture are available.

95. The agricultural education faculty confers at least annually with the appropriate campus library staff(s) to review holdings and acquisitions and to discuss other needs.

Evaluation

96. A continuous systematic evaluation and an annual review of the teacher education program is conducted with the assistance of the advisory committee state staff, employers, and graduates and is based upon stated objectives included in the long-range and short-range plans.

97. The Standards for Quality Programs in Agricultural/Agribusiness Education are used in evaluating teacher education programs.

Evaluation as a Continuing Process

Institutions of higher education undergo NCATE evaluations at least every 10 years and NASDTEC evaluations every 5 years for purposes of accreditation and certification. During the intervening years, state, university, and other kinds of evaluation may be required. Thus, evaluation is mandated for faculty members regularly, while the organizational structure is generally such that time and resources for self-study are provided only periodically or when necessary to meet deadlines. When preparation for a visiting team must be done hurriedly, the result is always poorer quality and a lesser degree of satisfaction than desired. Continuous and systematic evaluation is offered as a reasonable way of meeting the external evaluation team more fully prepared and avoiding undue stress on faculty and staff members and on resources.

It is recommended that the teacher education staff emphasize the evaluation of one major segment of teacher education in agriculture each year over a five-year period. In this mode, all aspects of the program would be assessed every five years. Additionally, evaluation would be continuous, and the teacher education program would be reasonably well prepared for an external evaluation. Gathering of data and the assessment of those data should be an integral, on-going part of the program operation.

A review of the several guidelines and standards presently being used suggests that the following five areas would encompass a teacher education in agriculture program.

1. Faculty. Teaching, scholarly activities, research, and professional activities.

2. Students/graduates. Recruitment, screening, enrollment, counseling, advisement, placement, and follow-up.

3. Curricula/programs. General, professional, and technical studies; field experiences, including student teaching; and leadership.

4. Facilities/resources. Classrooms and laboratories, libraries and library holdings, instructional media centers, and clerical and support staff.

5. Governance. Administration and organization, financing, and planning.

These five areas have no sequential order. One approach is to begin with the area that seems to need immediate attention. For example, it could be the area of faculty, if some staff members are being considered for promotion and/or tenure. It could be the area of curricula/programs, if the state has recently legislated the addition of courses of teaching reading and intercultural relations as requirements for teacher certification.

Faculty

The need to determine which personnel should be retained and/or promoted is the primary reason why agricultural teacher educators are evaluated. Faculty professional development, an important secondary purpose, is often given limited attention, except by the individuals personally involved and the department chairperson. Criteria for evaluating faculty positions and faculty persons usually consist of teaching, research accomplishments, professional activities, and community service. The weight given to each of the criteria measures defines the position or job description for a faculty member.

Because the teaching of teachers and prospective teachers is an art as well as a science, it is challenging to evaluate. Assessment should be done using a broad spectrum of both process and product data. The kinds of data which may be reviewed and analyzed include: (1) course outlines, teaching aids and examinations prepared by the instructor; (2) folders and records maintained on advisees and students; (3) papers, theses, and dissertations prepared by students under the direction of the instructor; (4) part-scores or grades made by students on subject matter areas of the comprehensive and qualifying examinations for which the instructor was charged with teaching; (5) feedback resulting from colleagues' observations of the instructor's teaching; and (6) students' feedback on courses taught by the instructor.

Tom and Cushman developed and field tested "The Cornell Diagnostic Observation and Reporting System for Student Description of College Teaching" (13.26). The rationale adopted for the study was that "a defensible measure of a student's achievement in a course is a self-rating measure of a student's achievement of progress in reaching objectives the instructor considers important." It was therefore assumed that teaching behaviors influence student achievement at the college level. The researchers concluded that "student self-ratings of progress on objectives valued by the instructor as well as their description of their instructor's teaching behavior are highly reliable." The Cornell system can be effective for (1) identifying the strengths and weaknesses of an instructor's teaching, (2) prescribing appropriate specific suggestions for remedying the weaknesses, and (3) supplying appropriate evidence concerning the instructor's effectiveness to those members of the faculty and administration involved in decisions concerning his/her promotion.

Students/Graduates

Agricultural education programs seek the best candidates for admission into their teacher education programs. The selection and recruitment of students revolves around evaluative criteria such as previous transcripts, standardized test scores, letters of recommendation, and personal interviews. Agricultural work experience, while still important, is no longer a requirement for admittance, rather it is a requirement for teacher certification. It can be obtained through some form of cooperative education while the student is in the teacher education program.

When a student is admitted into a teacher education program, it should be with the full understanding that the department will provide reasonable human resources for teaching, guidance, and counseling. The student must weigh the balance between variables such as the number and quality of students and the number and quality of staff persons available to provide student services.

Teacher education has not completed its responsibilities when its students graduate. Placement and follow-up of graduates are functions of teacher education in agriculture which are invaluable to the individuals who are involved and to the individuals who must evaluate the total program. While undergraduate teacher education in agriculture was established to prepare individuals to become teachers of agriculture, graduate inservice agricultural education is charged with upgrading teachers to positions of master teachers, researchers, and administrators.

The success and advancement of graduates in agricultural education and in other closely related agricultural occupations reflects directly on the teacher education programs of which they are products.

Follow-up studies of graduates are conducted for a number of reasons. The Department of Agricultural Education at The University of Arizona emphasized the occupational status and the salaries of its graduates in a six-year follow-up study (13.27). The Department of Agricultural Education at Virginia Polytechnic Institute and State University, in a five-year follow-up study of agricultural education graduates, suggested various means for improving recruitment and retention/efforts (13.28). In addition, follow-up studies of graduates can obtain data from graduates on other aspects of the program, such as services, student life, advisement, instruction, course evaluation, and curriculum.

Curricula/Programs

Curricula/programs in agricultural education continue to change and differ among institutions across the United States. Boucher, in a status study involving 78 institutions supporting teacher education in agriculture programs, found that "the major changes reported include internships, certification in a specific taxonomy of agriculture, pass-fail marks for student teaching, coordination or work experience with industry, more flexibility in certification requirements and more clinical experience prior to student teaching (13.29)." In the study, for example, 66 to 80 quarter-hours or 45 to 56 semester-hours of technical agriculture were required by a heavy concentration of departments. This kind of information could be used to establish guidelines and standards for future evaluations.

Competency-based teacher education, or performance-based teacher education, an alternative approach to teacher preparation, has been one of the most significant curricular innovations in teacher education in recent years. Elam's conceptual model of performance-based teacher education (Fig. 13.1) establishes essential elements, implied characteristics, and related desirable characteristics of this relatively new approach to teaching and learning (13.30). In a later publication, Elam confirms that there now appears to be general agreement that a teacher education program is performance based if it includes the essential elements (13.31).

The PBTE approach is being tried in many institutions to improve and strengthen teacher education programs. Internally, teacher education programs are using a variety of models and materials in search of answers. Externally, the programs are being evaluated as a means of certifying graduates for the teaching profession. The NASDTEC recognized the importance of performance-based teacher education programs when it established standards for evaluating them. These standards are as follows:

> Standard I. For each preparation program the institution shall develop and adopt an explicit statement of "program exit" competencies that relate to the entry-level professional role. These competencies must include all of the criteria implicit in the General Standards (3.4) and the Specific Standards (3.5). . . .
>
> Standard II. The institution shall provide a program design (1) relating the competencies (cited in Standard I above) to modules, subcourses, or courses, (2) listing the learning activities involved, and (3) specifying the assessment techniques used to verify the attainment of these competencies.
>
> Standard III. To determine program effectiveness, the institution shall formally assess follow-up data to determine the relationship between "exit" competencies and initial professional role

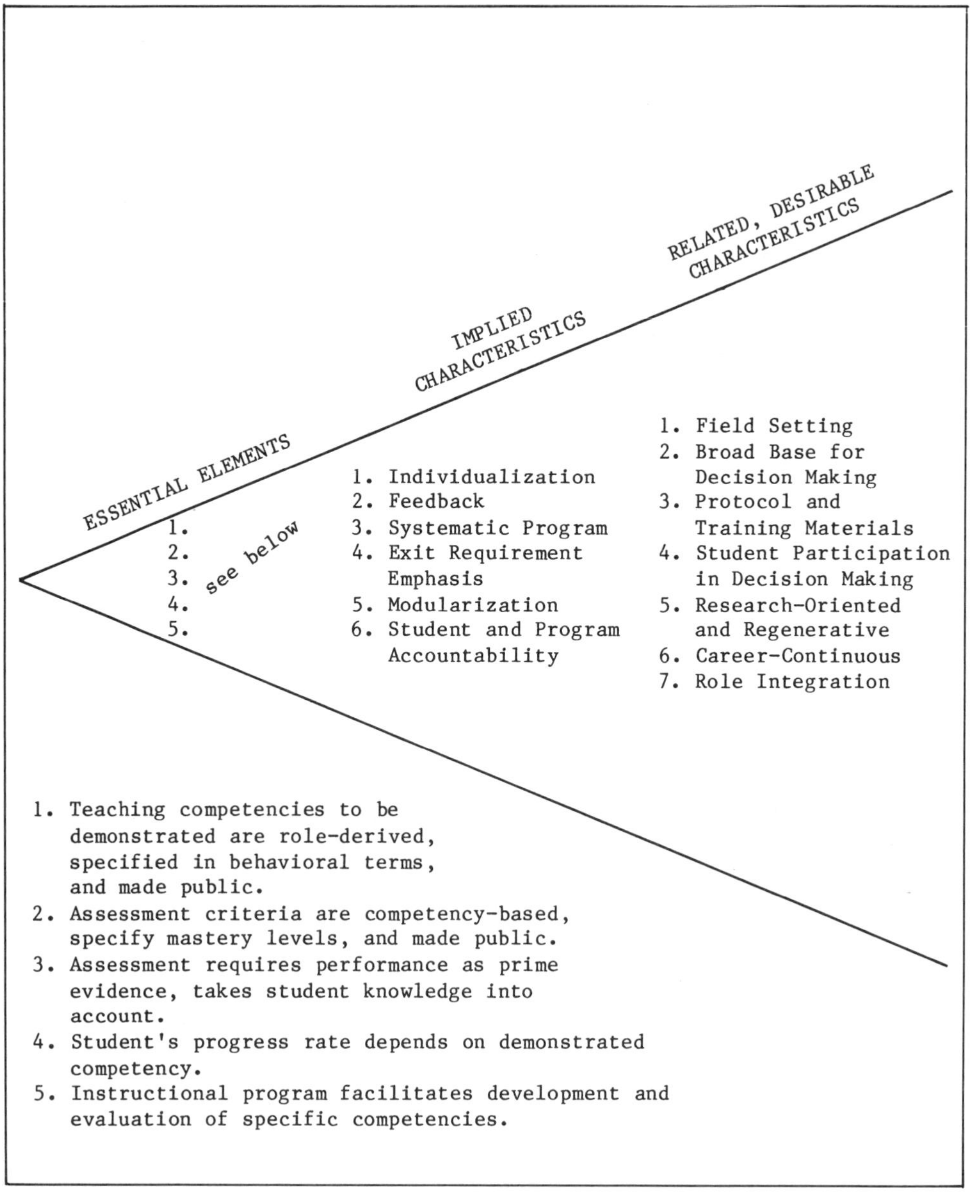

FIGURE 13.1. Conceptual model of performance-based teacher education, in Elam's Performance-based Teacher Education: What Is the State of the Art? (13.30).

performance. Such assessment should be the basis for continuous program development.

Standard IV. The institution's performance in the developing and verifying of a candidate's role competency and in collecting and using follow-up data shall be determined by an on-site evaluation team designated by the state education agency.

There is general agreement in the literature that evaluation and certification procedures are the major limitations of performance-based teacher education. If teacher education programs are to be certified on the basis of the demonstrated performance of prospective teachers, it follows that evaluative measures must be obtained through valid and reliable instrumentation universally accepted. Good teaching will need to be defined in a way that can be measured more objectively and that will be generally accepted. Also, collecting evidence of a candidate's ability to perform satisfactorily will be difficult to verify and costly to obtain. And, valid or not, it is likely that the responsibility for collecting evidence to support the certification of graduates to teach and the certification of programs to implement will be placed on the educational institutions.

Facilities/Resources

The kind and condition of classrooms and laboratories affect the quality of instruction which is provided and the amount of learning which results. Classrooms may range from rooms with four bare walls and chairs to carefully planned and adequately equipped modern demonstration instructional facilities. Agricultural laboratories may also fully support instruction, or they may be nonexistent. Equipment may vary from modern to obsolete in the mechanics laboratory; breeds may be limited in the livestock laboratory; visual equipment in the teaching laboratory may be difficult to schedule for use. Access to and the condition of classrooms and laboratories are criteria worthy of including in an instrument for assessing teacher education in agriculture.

Libraries need to be adequately staffed and well organized so that their resources can be used efficiently. The number of books and the other kinds of informational materials in the library are a quantitative means of evaluating it. The numbers and kinds of agricultural education resource materials in the library are a reflection on the agricultural education faculty members who have a responsibility for reviewing and continually recommending items for purchase by the library personnel. Library holdings at a university generally relate directly to the disciplines on campus which are using them. Therefore, agricultural education faculty members should periodically review the agricultural education holdings to maintain or exceed the standards established by the department or certifying agency.

An instructional media center is necessary for high quality agricultural education. There must be a direct correlation between the use given to instructional media centers and the quality of instruction.

Support personnel help the faculty to be more productive. The ratio of support staff to faculty is an important measure to use in evaluating agricultural education. Graduate assistants, work-study students, and secretarial staff members all aid the faculty. Not only are well-supported faculty members more productive, but they are usually happier with the end result of a high quality program.

In essence, education that takes place in attractive and supportive environments with a wealth of instructional materials and media managed by professionals who are aided by support staff is an asset to the learner. High quality programs have this kind of setting.

Governance

The data base for studying the organizational structure of teacher education in agriculture is practically nonexistent. Programs and departments are directly influenced by the university settings in which they function. Agricultural education programs and departments are located administratively in colleges of agriculture at some institutions and in colleges of education at others. In addition, each university has its own rules and regulations which affect the organizational structure of the department.

There are other factors which influence the organizational structure of agricultural education programs. Funding sources, state laws, certification requirements, and federal legislation dictate functions and assignments which relate to organization structure.

While the agricultural education programs at all institutions have similar missions and functions, each program is organized differently to meet its own goals. Both the organizational structure and its effectiveness in fulfilling the mission of the program should be evaluated.

Governance of a department is assessed in terms of accountability. The critical questions are (1) How was the money budgeted? and (2) What were the outcomes relative to graduates, research, and services? Facts and figures in the annual reports of the department can provide much of the accountability data necessary to determine the governance of the department, including the administrative and organizational structure, financing, and planning.

CHAPTER 14

Research in Teacher Education in Agriculture

Larry E. Miller
THE OHIO STATE UNIVERSITY

J. Robert Warmbrod
THE OHIO STATE UNIVERSITY

Planning, conducting, and disseminating research is an essential function of teacher education. The first section of this chapter includes the rationale for this premise and describes some specific responsibilities of teacher education for research in agricultural education. The second section is a description and an assessment of the status of research in agricultural education. Appraisals of research are reviewed; funding for research in agricultural education is discussed; and sources for locating research in agricultural education are presented. The third section addresses the quality of research in agricultural education. In this section strategies and procedures for improving the quality of research are given. The concluding section pertains specifically to research in teacher education in agriculture. Some of the major research in teacher education is reviewed, followed by proposals for additional research.

Research: A Function of Teacher Education

Teaching, research, and public service are the three major functions of public colleges and universities offering programs of teacher education in agriculture. Research is a major function of land-grant universities which account for three-fourths of the 84 public colleges and universities in the United States with teacher education in agriculture programs. The arguments for mandating research as a major function of academic units in a land-grant university apply, without exception, to agricultural education. Traditionally, most departments in colleges of agriculture have performed the research function proficiently through long-range planning for programmatic research, substantial and continuing financial support for research, faculty time specifically allocated for planning and conducting research, and systematic efforts to interpret and apply the research findings. Subsequent sections of this chapter will show that research in agricultural education is often described as lacking a programmatic emphasis and, in most universities, as having little if any continuing support from regular university, college, and experiment station funds.

Teacher education, particularly faculties in agricultural education in land-grant universities, has a clear mandate and responsibility to place high priority on research in agricultural education. The generation of new knowledge and the application of that knowledge to policy and program development are as crucial for the expansion and improvement of agricultural education as they are for other academic units of the university. Specific responsibilities for teacher education include identifying needed research in agricultural education, conducting research, disseminating and implementing research, and preparing researchers and consumers of research.

Identifying Needed Research in Agricultural Education

Agricultural education research periodically undergoes self-examination by members of the profession. Self-examinations vary from brainstorming sessions at regional research conferences to more formal attempts to identify and prioritize the concerns of the profession that need to be addressed. The prospective researcher will find a vast array of researchable topics, with each listing often reflecting the theme of a conference and/or the bias of the reporter for the review.

Agricultural education, like all areas of vocational education, has essentially assumed a reactive rather than a proactive posture on research. Funding procedures for research have necessitated this reactive position. Federal and state agencies have in essence established the research priorities of the profession, and researchers have had to respond to pre-established research needs if funding is to be obtained. Participation in the decision-making process for determining these needs has essentially not been shared by the profession. What started as closely prescribed grants has evolved into even more prescriptive contracts.

Nevertheless, agricultural education continues to conduct some research projects that address current needs, but these are often unfunded and are conducted at the sacrifice of funds for other departmental activities. However, optimism continues to prevail throughout the profession regarding the future for research activities.

In 1976, the Research Committee of the Agricultural Education Division of the American Vocational Association commissioned a study by Stewart, Richardson, and Shinn to investigate problems of the profession needing attention (14.1). The study sought opinions of supervisors, teachers, and teacher educators as to the most pressing problems. The data were analyzed according to geographic region, by groups, and in total. The problems identified provide a good basis for determining the perceived research needs in agricultural education. The areas of concern in rank order of importance were as follows: (1) curriculum development, (2) funding, (3) teacher education, (4) teacher shortage, (5) evaluation, (6) teacher certification, (7) supervision and administration, (8) adult education, (9) FFA, (10) manpower, (11) research, (12) post-secondary programs, (13) urban programs, and (14) administration. The top five items of concern under the third-ranked teacher education area are improving preservice programs, identifying competencies, requiring agricultural education versus education courses, requiring technical agriculture, and providing inservice education.

Conducting Research

Most research in agricultural education is conducted through departments of agricultural education. A consistently high percentage of this research is conducted through the efforts of graduate students (14.2). Drake (14.3) and Krebs (14.4) called for the development of a more programmatic approach to research. Krebs gives several suggestions for achieving a programmatic approach to research.

1. Commit the department to a programmatic research emphasis.
2. Select a department head who will be compatible with the research orientation of the department.
3. Recruit faculty members to fit into departmental research programs.
4. Recruit graduate students for specific research projects.
5. Develop a proper mix between basic and applied research.
6. Develop strong research relationships.
7. Teach graduate students and young faculty members the need to develop a lifetime research focus and commitment.
8. Analyze research completed to determine how findings were used.
9. Involve bright graduate students in research.

A survey by Warmbrod and Miller in 1979 showed some trends toward programmatic research in departments throughout the country (14.5). Forty-six land-grant universities responded to the survey, with 24 percent reporting that a position paper had been developed describing research efforts for their departments, and with 40 percent indicating they had a faculty member designated as a coordinator of research. Fifty-seven percent annually identified and conducted staff studies, and 65 percent reported that they set research priorities for the department.

The major responsibility for the conduct of research in agricultural education will undoubtedly continue to lie with teacher education. The land-grant philosophy provides for public service, teaching, and research. However, the increased research expertise of supervisory personnel and teachers may nurture others to also enter the field. Additionally, agencies outside the educational community are also competitively bidding for the larger grants and contracts, and this trend may continue. In 1979, Patterson supported research by units separate from a university system (14.6). He presents these major premises to his argument: (1) agricultural education is fragmented across colleges of agriculture and education; (2) the profession self-identifies research as less than a primary function of departments; and (3) research is often not a function in non-land-grant institutions. He further argues that research coordinating units in vocational education are now viable and can coordinate numerous agencies for the conduct, development, and utilization of research and that increased specialization for research expertise may best be developed outside the academic setting.

The conduct of research should be a joint effort of all within the profession. The identification of problems, the prioritization of concerns, the assistance with data gathering, and the dissemination and utilization of research results call for a coordinated effort that cannot be facilitated by any one group.

While teacher educators must assume a major role in the conduct of research, Warmbrod and Miller found that the ranked departmental emphasis were (1) teaching, (2) student advisement, (3) off-campus inservice education, and (4) research. One would be in a tenuous position in attempting to argue against the propriety of this rank order. However, aspiring teacher educators should be cognizant of the place research has in the activities of most departments. In 1980, Bowen reported teacher educators in agriculture devoted an average of 11.9 percent of their time to research, with a range from 1 to 65 percent (14.7).

Disseminating and Implementing Research

Dissemination and implementation of the results of research are important considerations for agricultural education at all levels. Dissemination has typically taken place through theses and dissertations, research reports, monographs, paper presentations, and journals such as The Journal of the American Association of Teacher Educators in Agriculture, The Journal of Vocational Education Research, and The Agricultural Education Magazine. The principal settings for presentation of research papers are the regional research conferences and the national agricultural education research meeting which are held annually. Implementation of the results of research is one of the most often criticized facets of the research process. The contention is often made that what is learned through research is not disseminated and put into practice. Regional and national conferences for agricultural education, including those specifically addressing research, provide an excellent means of involving teachers, teacher educators, and supervisory personnel in the dissemination of research. Dissemination activities that will stimulate the implementation of research findings should continue to be encouraged.

Because agricultural educators continue to be their own worst critics, this may affect dissemination. A mythical standard seems to exist against which research is compared. Often suggestions for improvement are made so that agricultural education research will be more widely recognized and accepted by the "research community." There is always room for improvement. However, self-flagellation can be carried to an extreme. The profession needs to be no more self-critical than any other discipline. In fact, research in agricultural education has made significant contributions to the field, with perhaps as great a success rate as in any other profession or field. Certainly, continuous improvement should be sought, and perhaps referees, discussants, and peers are sometimes too kind in reviewing the work of others. A proper mix of the two extremes needs to be obtained.

Preparing Researchers and Consumers of Research

Secondary and post-secondary teachers of agriculture should be intelligent consumers of research. A major portion of what they teach, that is, the science and technology of agriculture, is the result of research in agriculture. To be effective, teachers must be able to search out relevant agricultural research, interpret and summarize the research accurately, and communicate the salient findings clearly and concisely. In addition, they must possess a high level of competence and skill in locating and using research in agricultural education if they are to increase their proficiency in planning, conducting, and evaluating programs of instruction. A major responsibility of teacher education, both in preservice and inservice programs, is the preparation and continuing development of present and prospective teachers to be competent consumers of research.

A major element of graduate education in agricultural education is the development of knowledge and skill in the planning and conducting of research. A 1980 study by Osborne showed that graduate education at the master's degree level required the preparation of a thesis or the completion of a research project, a developmental project, or some form of independent study that required knowledge and skill in the use of disciplined inquiry (14.8). The rationale for this requirement is that proficient agricultural educators must be able to identify and define significant issues that need to be investigated and to design and conduct systematic studies that yield valid and reliable data.

The preparation of researchers is a major part of advanced graduate programs in agricultural education. Those earning the doctoral degree with major specialization in agricultural education are the primary researchers in agricultural education. If the significant issues in agricultural education are to be researched in a systematic and expert manner, the competence and skill required must be acquired through graduate programs offered by faculties in agricultural education.

Status of Research in Agricultural Education

Who Conducts Research in Agricultural Education?

Earlier reference has noted that most research in agricultural education is conducted in teacher education, and the majority is done by graduate students. Through analyzing summaries of studies for 1974-1977, Mannebach noted that 57 percent of the research reported was accounted for by the states of Ohio, Oklahoma, Pennsylvania, Iowa, and Virginia (14.9). He also indicated that 51 percent of the studies summarized were master's studies, 27 percent were doctoral research, and 22 percent were staff studies. Seventy-two percent of the research was reported by the Central and Southern regions.

Characteristics of Research in Agricultural Education

Krebs provides insight into the characteristics of research in agricultural education. His 10 points are presented as a means of self-examination by the profession and should receive due consideration (14.10).

There is a tendency toward fragmented one-shot studies. Just for kicks, I asked several agricultural educators to name colleagues noted for research on specific topics in agricultural education. Despite some prompting, there were no responses--no persons identified with research in particular problems. This is an expected result from a field in which individuals have not spent much of their research time on specific research problems over a professional lifetime. A program is researched once and the job is considered done for all time.

Agricultural education research is dominated by surveys. This point was not really given as a criticism, just as a statement of fact. Whether this is true or should be true is for you to judge.

Studies lack depth. Agricultural educators have a tendency to "skim" a problem rather than study it in depth. I might add that statistical treatment of data does not, by itself, change that impression.

Results are reported and applied before fully supported by research. This was a rather peculiar comment intended to convey the thought that agricultural educators tend to make premature application of findings--application of findings not yet fully supported by research. There may be too much of a rush to get into print, or the hurry may be pressure from sources of funding. Testing is not as intense as in other disciplines.

Funding is the basic orientation for planning. Planning based on the vagaries of federal priorities is not a wise policy. This is, of course, easy to say for those in areas with no funding problems. However, long term planning must be based on long term objectives.

Research program development is non-existent. This point supports the first point made. In fact, it appears there has not even been an identification of the areas in which research impact is desired. Certainly, there is no strong identification of individuals with specific areas of research over a long period of time.

Research is not fully focused on agricultural education. This may be one of the reasons why agricultural education research hasn't gained more attention.

Confusion exists between research and development. Development projects are often called research or at the very least research and development. This may, in fact, be the basis or cause of the earlier comment regarding premature application of fundings. A funded project is a funded project and a research identification seems more scholarly than a developmental identification. Institutional emphasis in evaluating faculty is also important here.

Agricultural education research is limited in applicability to other geographical areas in agricultural education and to other fields of study. While this was given as a characteristic, the point could be made that the studies were better than the use made of the results. Even some of the long-time follow-up studies did little to change the image of research in agricultural education as just described. While the results of placements in jobs could and should have changed the publicly stated objectives of vocational agriculture programs, the fact is that it took federal legislation to point the program in the direction indicated by the research.

Findings are not applied. Much research fills shelves and collects dust.

Reviews of Research in Agricultural Education

Two major publications summarize published and unpublished research in agricultural education. Agricultural Education: Review and Synthesis of the Research by Newcomb is an example of a 10-year review of research in the field. The latest volume, published by the ERIC Clearinghouse on Adult, Career, and Vocational Education, the National Center for Research in Vocational Education, reviews research studies conducted between 1969 and 1978 (14.11). Similar reviews of research in agricultural education were authored by Warmbrod and Phipps (14.12) in 1966 and Carpenter and Rodgers (14.13) in 1970.

The second publication is the Summaries of Research and Development Activities in Agricultural Education. This summary is compiled annually through the Research Committee of the Agricultural Education Division of the American Vocational Association. It provides a state-by-state listing of research and development activities completed in a given year. Studies in progress are also reported by state and indexes crossreference the summary. The recent editions also contain ERIC codes to assist in locating other summaries of studies in agricultural education. The Annual Review of Research in Vocational Education, Volume 2, published by the Department of Vocational and Technical Education at the University of Illinois, also contains a review of research in agricultural education. Additionally, potential researchers will find it advantageous to search the ERIC system, perhaps through a computerized search, to locate appropriate research. Computerized searches can also be used to locate agricultural related subjects (AGRICOLA), cooperative state research of the USDA (CRIS), and the comprehensive dissertation index. A search of foundations as funding sources may be done through the Foundation Directory, Foundation Grant Index, Frost and Sullivan Defense Market (DM) measures system, national foundations, Grants Database, Smithsonian Science Information Exchange (SSIE), and current research.

Support for Research in Agricultural Education

Agricultural education research draws its support from a variety of sources. Many of the studies by teacher education faculties are supported through departmental budgets. Numerous studies conducted by graduate students are self-financed.

Exterior funding comes from a variety of sources. Warmbrod and Miller found that nearly 52 percent of the funds for research were generated through responding to requests for proposals (RFP's) (14.14). Evidently, very few departments have line items in their budgets for research activities. Sources of funds reported by the departments of agricultural education for 1979-80 are shown in Table 14.1.

TABLE 14.1--Sources of 1979-80 research funds for departments of agricultural education

Source	Number of Departments Reporting	Average % of Funds from Source/Range[a]
State departments of education	35	69.2 / 35-100
Agricultural experiment stations	15	37.4 / 5-80
Universities	16	31.3 / 5-100
Federal agencies	7	28.1 / 10-75
Foundations	2	19.0 / 0-35
Other	5	18.3 / 10-25

[a]Percentages do not total 100, since the average reported is the mean percentage of those reporting support from that source.

State departments of education provide the highest average support, with 35 departments out of 46 reporting some funds received from this source which supplied an average of 69 percent of the funds. Sixteen departments reported an average of 0.67 full-time equivalent (FTE) faculty positions, with 1.33 graduate student positions funded from this source.

Fifteen departments reported obtaining support from agricultural experiment stations. For these departments, this support averaged 37 percent of their research funds. Eleven schools averaged 0.67 FTE faculty positions, 1.33 graduate students, and 0.5 clerical positions funded, with 14 departments also receiving money for travel and supplies.

Support for research from within the university was reported by 16 departments, and this accounted for an average of over 31 percent of the research budgets of these departments. Five schools averaged 0.6 FTE faculty positions funded through university research funds, four reported one graduate student funded, and three averaged one-third of a clerical position. Eight reported university research monies for travel and supplies.

Seven departments reported receiving funds for the conduct of research from federal agencies. These funds provided an average of 0.57 FTE faculty positions for four departments, four departments averaged 1.5 graduate students funded, and six had approximately one-third of a clerical position funded, as well as funds provided for travel and supplies.

Foundations and other agencies provided research funds to five departments. Three departments averaged one-third FTE faculty positions. Five departments reported an average of 1.70 graduate student positions and 0.63 clerical positions from these funds. Eight departments reported receiving research funds for travel and supplies from these sources. The principal source reported under this category was directly from agribusinesses and not necessarily from the foundations supported by these agribusinesses.

Funds to support research in agricultural education are increasingly difficult to obtain. At the federal level, most research funds have been withdrawn from the particular bureaus and placed within the National Institute of Education (NIE). The NIE has been beset with difficulties from within and without. Serious problems of changes in leadership personnel and direction have limited the contributions of the NIE to all fields of vocational education. Funds for research through vocational legislation at the federal level are principally allocated directly to states. The states vary greatly in how they utilize these funds. In some states most of these funds are used for personnel to conduct enrollment and follow-up studies of vocational education students for federal reporting. In some states where funds are allocated for in-state RFP's, research funds are often used for curricular and other developmental activities instead of research activities. Research is interpreted in these states to connote data procurement for federal reporting or developmental activities. Arrington's review of fiscal year 1978 and fiscal year 1979 state program improvement undertakings indicate that during these two years, $39,205,436 was expended on 1,560 projects across all the states, the District of Columbia, and Puerto Rico (14.15, 14.16). Projects related to agricultural education accounted for only 83 of these projects and a total of $1,902,119, which was 4.9 percent of the funds. Further analysis of projects according to problem areas addressed revealed that 35 projects in agricultural education were in curriculum management/development, accounting for 52 percent of the funds expended, and that 8 were for special needs, with 10.7 percent of the funds. The remaining funds were under the headings of urban/rural youth;

equity/civil rights; education-to-work transitions; planning, data, and accountability and adult education. These categories varied from 5.9 to 1.2 percent in accounting for the funds utilized by agricultural education. Information was unavailable for nearly 23 percent of the projects. The data also indicated that in agricultural education, 65.3 percent of the improvement activities are conducted by four-year colleges and universities, 16.6 percent by local educational agencies, 6.7 percent by two-year institutions, and the remaining 11.4 percent by state agencies and private business.

Procurement of funds through agricultural experiment stations seems to be a growing trend for departments of agricultural education. Those departments administered through colleges of agriculture may find this source of particular interest.

Warmbrod and Miller requested departments to indicate the type of research they conducted. Thirty-three percent characterized their research as applied, 31 percent as developmental, 36 percent as curriculum material development and testing, and 0 percent as basic research (14.17). While the definition of types of research was left to the respondent, the nature of research in agricultural education is clear in one regard--basic research is not considered a type of research conducted.

Procuring funds has become important to most departments of agricultural education. Therein may lie the rationale for the reporting of the type of research conducted. Funds secured for the development of curriculum materials, for the field testing of these materials, or for the conduct of professional development activities may be most easily administered through research accounts. Nonetheless, money procurement (obtaining grants and contracts) has become an essential aspect of departmental operation. Not only are grants and contracts essential for research, but the overhead from indirect costs helps maintain the on-going operation of the department. Many departments would be hard pressed to offer assistantships to graduate students without the benefits derived from external funding.

The rewards in the higher education community of salary, rank, and tenure have traditionally been based upon research publications, teaching, and field service. The priorities for receiving the rewards are usually reflected in the order of the preceding listing. With finances for higher education becoming more restricted and retrenchment occurring, the ability to obtain external funds will become increasingly important. The ability to procure external funds may even eventually exceed the traditional determiners of rewards. A grantsperson able to procure, for example, $100,000 per year, with 40 to 60 percent indirect costs becomes a valuable person to an institution of higher education. Experience in the writing of proposals should constitute a segment of the preparation of any graduate student aspiring to a leadership role in agricultural education.

Support for research also needs to be inherent in the philosophy of the departments. Time and financial resources need to be allocated by departments to ensure that faculty members are able to prepare proposals and to conduct research. Too frequently research is the bottom rung on the ladder of departmental priorities; yet, it is the top rung of criteria for rewards. Thus, a paradox exists. Each professional must resolve this enigma for himself/herself. If research continues to overload the faculty, it will not continue to be of high quality. Administrators must work cooperatively with faculty to clear up this dilemma for junior faculty members seeking promotion. Lofty expectations for faculty require lofty support.

Improving Research in Agricultural Education

A major goal of teacher education is the continuing improvement of research in agricultural education. To achieve this goal, teacher educators must consider the soundness, the rigor, and the degree of sophistication with which research is planned, executed, and reported. Disciplined inquiry--the method of science--must be executed expertly if high quality research is to result. To achieve high quality research, teacher educator must accomplish each of the following phases, or steps, of the research process in a highly expert manner: (1) selecting and defining the problem and developing specific objectives, (2) accumulating pertinent knowledge and information about the problem being investigated, (3) designing the study and collecting data, (4) analyzing data and drawing conclusions, and (5) reporting research. The comments that follow highlight factors that should be considered for each step in the process if the goal of improved research in agricultural education is to be achieved.

Selecting and Defining the Problem to Be Investigated

The most essential indicator of high quality research is the nature of the problem being investigated. The late H. M. Hamlin in an article titled "What Is Research?," published in the September 1966 issue of the American Vocational Journal, stated this criterion succinctly: "Ideas and concepts are the first requisite for good research. . . . Research without important ideas is busy work." High quality research in agricultural education must begin with major ideas, issues, and problems. Research problems are significant when their solutions lead to the generation and expansion of knowledge that is useful in policy and program development in agricultural education.

Much of the research in agricultural education is decision-oriented research. The primary end sought through the research is the generation of knowledge that contributes directly to policy formulation and program development, revision, and improvement. Comments that imply that high quality research is possible only if a certain methodology is used, namely experimentation, are not uncommon. That point of view is a narrow and inaccurate perception of both the nature and the purpose of the research process. The first test that should be met in making decisions about problems and issues to be investigated is determining the substantive ideas and concepts basic to the research, not the design and methodology required to conduct a valid study. Decisions about the significance of possible research problems in a given area of study are largely judgmental and best made by persons with a great deal of expertise in the area in which the research is being conducted.

The lack of programmatic emphasis in research on agricultural education can be responded to through careful attention given to the selection of research problems. Krebs (14.18), Taylor (14.19), Brown (14.20), and Copa (14.21) discuss the advantages of programmatic research and suggest strategies for long-range planning toward a more programmatic emphasis for research in agricultural education.

Accumulating Pertinent Knowledge and Information

Research in agricultural education will not be rated as high quality if it is evident that the researcher is uninformed about the major ideas, concepts, and theoretical constructs basic to the research. Expert use of the research process requires that the investigation of important problems proceed in light of what is already known.

A thorough knowledge of the theory and research pertinent to the problem being investigated is essential. First, related research and theory enable the researcher to identify and define significant research problems, including the identification of important variables that must be investigated. Second, theory and related research provide the rationale for research hypotheses that will be investigated. Third, related theory and research provide a framework within which the findings of the investigation can be interpreted. And fourth, the findings of research that are undergirded with theory and related research are more likely to be cumulative; that is, the findings will contribute to an expanding knowledge base in agricultural education.

A legitimate criticism of research in agricultural education is that it is frequently conducted in isolated, fragmented segments, with little attempt made to relate what is being researched to what is already known. More attention to accumulating pertinent knowledge and information can contribute to the alleviation of this criticism. Copa maintains that reviews of literature and professional reading required for conducting research in vocational education must extend outside vocational education, drawing upon the knowledge base in psychology, sociology, economics, and philosophy (14.22). Copa argues that researchers in vocational education must become information brokers who bring together the problems of those who practice vocational education and the various knowledge bases in the traditional disciplines.

Designing the Study and Collecting Data

If valid data and information are to be the outcomes of research, the researcher's first and foremost design consideration must be the interpretability of data. The study must be designed to yield data that are valid. The researcher's first concern must be the internal validity of the design. In exploratory and descriptive survey research, internal validity takes the form of valid and reliable measurement of variables, which means that measurement error is minimized. In ex post facto and correlational research, internal validity usually is very complex, since the goal is to maximize the possibility that relevant independent variables are being investigated and interpreted properly as factors explaining the variability in the dependent variables. In experimental research, internal validity is very important in that it deals with the basic question of whether the treatment or presumed cause has in fact produced the effect that has been observed.

If the design is to yield valid data, the researcher's tasks are clear. First, possible sources of invalidity must be anticipated and recognized. Second, the design and conduct of the study must be such that the real and anticipated threats to internal validity are minimized or hopefully alleviated. Finally, the consequences of uncontrolled threats to valid interpretation are recognized and made explicit. Brown offers helpful suggestions for dealing with two major design issues--collecting data and identifying and controlling extraneous variables (14.23).

Once the internal validity of the data has been maximized, the researcher's second major design concern is generalizability, or external validity. Depending on the nature of the problem and other factors, the researcher may or may not wish to extrapolate the findings of the research to groups other than those involved in the research or to situations and environments other than those in which the study was conducted. As in the case of internal validity, a mark of high quality research occurs when the researcher's intent regarding generalizability is made clear and is accompanied by design strategies that allow the degree of extrapolation desired.

If generalization to a target population is intended, sampling is a major design consideration. Sometimes the researcher overlooks or ignores that it is one thing to draw a representative or random sample, but it is another thing to deliver a representative sample. Since much research in agricultural education is descriptive, relying heavily on mail questionnaires, it is also important that relevant external validity design considerations not be ignored.

Analyzing Data and Drawing Conclusions

Analysis of data and the design of research are highly interdependent. The design of research, in conjunction with the nature of the data, fairly well prescribes the range of analysis techniques from which the researcher can choose. Sometimes the implication is made that esoteric and sophisticated statistical analyses can be used to rescue a narrow, insignificant research concern that has been investigated in an amateurish manner. However, complicated and sophisticated statistics, whether appropriately or inappropriately used, do not guarantee high quality research. Relying on statistics alone to signal the credibility of research is effective only with those who are uninformed and consequently awed with statistics and with those who are too timid to ask questions about the analysis techniques used.

When research is competently conducted and reported, it is evident that the researcher has used statistics properly as a research tool. Helpful guides for selecting appropriate analysis techniques are provided by Oliver (14.24), Oliver and Hinkle (14.25), and Brown (14.26). Researchers in agricultural education must become more adept in the use of multivariate techniques. The expert researcher keeps in mind that statistics is a tool to aid in decision making. Statistical analyses do not reveal what decisions the researcher must make; statistical procedures only provide information, both descriptive and inferential, that can be used in interpreting data and in making generalizations about the applicability of findings. Through the use of appropriate analysis techniques, the researcher demonstrates expertise in executing high quality research by developing interpretations that are substantiated by valid data and by formulating generalizations that are warranted by the design and conduct of the study.

Reporting Research

The research process is not complete until findings are reported. The research process, or the method of science, is an open and public process. So a final criterion of high quality research is the manner in which the research is reported. Researchers do not proclaim their research to be of high quality. That label can be bestowed only by their colleagues and peers. Consequently, research that is never reported has no chance to be considered of significance and high quality. Information regarding the reporting of research is offered by Warmbrod (14.27), Copa (14.28), and Miller (14.29).

Other Important Factors

Resources for Research in Agricultural Education. Significant further improvement in the scope and quality of research in agricultural education will require that substantial funds that are earmarked specifically for research be allocated on a continuing basis. The survey of departments and faculties of agricultural education in land-grant universities reported earlier indicates that some progress is being made in securing continuing support for research in agricultural education, particularly through agricultural experiment stations. Funds must also be sought aggressively from state and federal government agencies and from foundations with special interest in agricultural education and vocational education. Within universities, funds need to be allocated specifically to support research in agricultural education to include faculty positions for persons whose major responsibilities are to plan and conduct research.

Cooperative Research Projects. Agricultural educators, particularly teacher educators, can make significant improvement in the contribution of research to policy and program development in agricultural education through cooperative research projects. The regional conferences on agricultural education provide the mechanism whereby a number of states can cooperatively plan and conduct research on significant problems in agricultural education. Such a pooling of talent and resources will enable indepth research that has broad generalizability to be well planned and conducted.

Research in Teacher Education in Agriculture

Major Topics and Issues Being Researched

Research in agricultural education is broadly diverse. Newcomb notes that one of the frustrations of preparing the review of research conducted between 1969 and 1978 was trying to establish a typology for the larger number of areas investigated (14.30). His review of teacher education focuses upon competency-based teacher education, analyses and evaluation of teacher education programs, student teaching, problems of first-year teachers, and miscellaneous studies. As Krebs points out to teacher educators in agriculture, departmental and professional research is not very programmatic (14.31). The study by Bowen is an attempt to look at the job satisfaction of teacher educators in agriculture, and it may serve as a precursor for other studies directed to examine the profession using other variables (14.32).

The previously cited study by Stewart, Richardson, and Shinn identified some of the problems in the profession that need attention. The rank order of the areas of concerns identified has some similarity to studies conducted and categorized by Newcomb. The dissimilarity in the listings sheds light on areas needing further research.

Research Needed in Teacher Education

Newcomb asserts that research in agricultural education is not what it should be and calls for quality research in the basic tenets of agricultural education (14.33). He notes that the quality of research is improving but that there is room for improvement. There is a plethora of potentially researchable topics yet to be investigated. Additionally, replication of studies should be more fully considered, as many studies are limited in their generalizability to specific areas or states, or to particular ecological circumstances.

Copa challenges all areas of vocational education to investigate three basic categories if research is to progress. His first category is "meaning," where he seeks common definitions to terms that are too often operationally defined for a specific study. Second, he asks the profession to look at "ends," wherein researchers investigate the purposes served by vocational education and how such can be met. His third category is "ways and means" which addresses how to best get the job done. He maintains that there has been an overemphasis on means, that research skills have been restricted to empirical methodology, that there are restraints to linking research to practice, and that there is little programmatic research (14.34).

Brown advocates improvements in research in agricultural education in the areas of (1) selecting research problems, (2) developing data collection techniques, (3) selecting data analysis techniques, and (4) identifying and controlling extraneous/error variances (14.35).

Copa and Brown both propose strategies for improving research. Inherent in their recommendations are potential topics and ways to guide future research that would provide guidance for new research efforts or the replication of previous research.

Numerous questions in the profession need to be answered. These questions should have a solid theoretical framework. Copa's categories can provide such a framework to guide the profession in using sound methodology to address some important problems and issues.

Agricultural education is blessed with professionals who have outstanding skills in research. However, a substantial number of studies could be benefited if teacher education sought assistance from other disciplines and other experts within, and outside, the structure of the colleges and universities involved. Many questions might best be answered by national studies or consortia arrangements between states and institutions.

Summary

Planning, conducting, and disseminating research in agricultural education is a major function of teacher education. The generation of new knowledge and the application of that knowledge to policy and program development is as essential for the expansion and improvement of agricultural education as it is for all specialized fields of endeavor in education and agriculture. Specific responsibilities for teacher education include identifying needed research in agricultural education, conducting research, disseminating and implementing research findings, and preparing researchers and consumers of research.

Most research in agricultural education is conducted by units in colleges and universities charged with the responsibility for preservice and inservice teacher education in agriculture. Most research is done by graduate students, particularly master's degree candidates. Research in agricultural education is largely descriptive, often lacking a long-range programmatic emphasis. There is not substantial or continuing funding for research in agricultural education. Considering the limited support for research in agricultural education and the high demands for teaching and public service of teacher educators, the extensiveness of research in agricultural education is noteworthy. Dissemination of research in agricultural education is accomplished primarily through the annual *Summaries of Research and Development Activities in Agricultural Education*, the annual National Agricultural Education Research meeting, and *The Journal of the American Association of Teacher Educators in Agriculture*.

Teacher educators should continue to place high priority on the improvement of research in agricultural education. The expert use of disciplined inquiry requires a high level of competence in selecting and defining problems, reviewing related research and theory, designing valid studies, analyzing and interpreting data, and reporting research. The need for additional resources to support research and the advantages of cooperative research projects are proposed as factors essential to the further development of research in agricultural education.

CHAPTER 15

The Role of Teacher Education in International Agriculture

William L. Thuemmel
UNIVERSITY OF MASSACHUSETTS

Donald E. McCreight
UNIVERSITY OF RHODE ISLAND

Richard F. Welton
KANSAS STATE UNIVERSITY

A shrinking world and increasing global interdependence are bringing new challenges and opportunities to teacher educators in agriculture. Organizations involved with international agricultural and rural development efforts are placing increased emphasis on technical assistance and training related to improved and appropriate methods for the transfer of technology. A marked shift has also occurred in recent years to training programs which are more people oriented. Consequently, teachers of agriculture, extensionists, and related human development specialists are now sought for overseas development assignments on a scale commensurate with that experienced by agricultural scientists and technicians in recent years.

During the past two decades, agricultural institutions in the United States have been increasingly called upon to expand their participation in international development activities. This is especially true among the developing, or less developed, countries (LDC's). Agricultural educators possess a basic knowledge of applied agricultural science, pedagogical skills, and a vocational orientation. Thus, they have much to contribute as professionals to international agricultural and rural development. However, very little information has been available regarding the extent to which U.S. teacher educators in agriculture have been involved, or even interested, in international activities.

Thus, this chapter provides a background and rationale for involvement by teacher educators in international agriculture and describes the current status of agricultural teacher education participation in international activities. To assess the current state of the art, Thuemmel and Welton conducted a nationwide survey of agricultural teacher education activity in international agriculture in 1979 (hereafter referred to as the AATEA Survey) (15.1). Survey data on teacher education international activities were obtained both on institutional involvement and individual teacher educator interest and experience. Much of this information is presented and discussed in describing the role of agricultural teacher education in international agriculture.

Rationale for Professional Involvement in International Agriculture

Agricultural education has evolved basically as a profession built around a federally sponsored program to prepare teachers of vocational agriculture. Since passage of the Smith-Hughes Act in 1917, departments or divisions of agricultural education have been a part of most land-grant universities. The Vocational Education Act of 1963 and the Vocational Amendments of 1968 have broadened the scope of vocational agriculture. However, for the most part, agricultural education is still viewed by many as having one major function--namely, the preparation of secondary teachers of vocational agriculture. Bender, writing in 1977, believes "that the programs of agricultural education have been limited too much to this single function (15.2)." He referred to a statement prepared in 1967 by the National Science Foundation-sponsored Commission on Education in Agriculture and Natural Resources (Committee on Agricultural Education co-chaired by Kottman and Metcalfe) concerning agricultural education for the 1970's and beyond. This report forecasted a much broader role for agricultural education, noting that:

> The responsibilities of staff members in agricultural education will extend to encompass some functions not now regularly carried out. They will establish contact with a wide clientele. Regardless of the form that teacher education programs take, the underlying purpose is to develop understanding and an ability to design, implement and evaluate educational programs (15.3).

The committee's report further indicates that if teacher educators are to meet their responsibilities as staff members, they will need to design curricula for a wide range of clientele including "agricultural educators who will be entering international education programs (15.4)."

Teacher educators could well be at the crossroads in establishing their commitment as a profession to agricultural and rural development outside the United States. Since the early 1950's, much of the agricultural assistance granted to LDC's has been in the form of improved technology to increase production. However, it is now realized that the simple transfer of existing technology, managerial skills, or capital resources does not necessarily provide adequate or desirable solutions. Cultural adaptation and new approaches are recognized as essential factors in providing people-oriented programs that will benefit the rural poor in the LDC's. Many teacher educators in agriculture have the necessary skills and motivation to play key leadership roles in developing and implementing new training programs for personnel involved in international agricultural and rural development. This new approach could effectively provide for the transfer of technology in a manner which is more acceptable and adaptable to rural populations than in the past.

Legislation and International Agricultural Education

Agricultural teacher educators initiated the 1980's with a heightened self-awareness about their professional worth and relevance in the field of international development. Much of this positive self-realization in the agricultural education profession has resulted from emerging international activity among many colleges of agriculture in the United States under Title XII of the International Development and Food Assistance Act of 1975 (Public Law 94-161).

Title XII funding has shifted the focus of U.S. foreign assistance efforts from a broad basis to helping the poor majorities of the most poverty-stricken nations--"the poorest of the world's poor." This 1975 federal legislation amended the Foreign Assistance Act of 1961, which created the U.S. Agency for International Development (USAID) to administer all economic and technical assistance overseas programs. Title XII also gives the land-grant and other eligible universities a greater role in the planning, management, and evaluation of foreign assistance projects. In fact, Title XII strengthens the capacities of the eligible universities--many of which have agricultural education faculties--to participate with USAID as equal partners in foreign agricultural and rural development efforts.

Universities and land-grant colleges are now challenged by a mission that extends beyond the campus, state, and nation to one in which their expertise and resources can be of service to all humankind (15.5). However, some agricultural development specialists see foreign assistance efforts by U.S. universities as being too sophisticated for the needs of many less-developed countries. They believe agricultural educators could, and perhaps should, play a more active and effective role in international agriculture. During the mid-1960's, Hannah observed:

> . . . emerging developing countries do not need the professional level and research development that is presently the program of most of our land-grant colleges. What is really needed is the initial philosophy of these colleges when they were established by the Morrill Act. What is needed is a system of education that provides basic know-how about production (15.6).

Matteson reported that Title XII has "placed universities in a somewhat strange, if not awkward, position" because "historically, these institutions have not had extended involvement in the development and conduct of programs for the poor majorities of the developing countries." He believes "the greatest reserve of personnel which have experience working with small farmers, and in many cases poor farmers, rests within the ranks of the Agricultural and Extension Education profession (15.7)."

Thuemmel (15.8) has echoed the views of both Hannah and Matteson. However, neither teacher educators nor rich nations have all the answers for the agricultural and rural development problems of the less-developed countries. The challenge for improving the living conditions of the world's poor demands continued international cooperation and professional teamwork among the various technical assistance organizations and development specialists. Title XII has provided U.S. universities and agricultural educators with a new mechanism for actively responding to this challenge.

Institutional Involvement

Most universities in the United States are visited each year by many foreign scholars, scientists, government officials, and teachers to exchange ideas and knowledge. These visits may vary in length from one day to a year or more. Each year many American professors and students travel abroad to enrich their cultural dimensions. However, according to the AATEA survey, only one-half (50 percent) of the agricultural teacher education programs in the United States were reported to have been involved in international activities, either at home or abroad, during the past two decades. The survey obtained data on agricultural teacher education participation in international activities conducted either on U.S. campuses or at overseas locations between 1960 and 1979.

Involvement on U.S. Campuses

Nearly one-half (48 percent) of the 84 agricultural education programs or departments in the United States responding to the AATEA survey indicated they have been involved with international activities on their university campuses during the 1960's and 1970's. On-campus international activities were most prevalent in those agricultural education programs situated in the central (72 percent) and eastern states (69 percent). In contrast, programs involved with on-campus international activities were in the minority on university campuses in the southern (36 percent) and western (29 percent) states. Three times as many on-campus international activities were reported for the 1970's as for the 1960's. During the past 20 years, those activities have been reported to have involved the part-time services and expertise of nearly 100 different teacher educators.

The nature of international activities involving agricultural teacher education faculties varies greatly from one university to another. On most campuses (70 percent of those reporting international involvement in the AATEA survey), international activities are limited to advising and teaching foreign students enrolled in domestically oriented agricultural education undergraduate and graduate programs or in selected adult, extension, or teacher education courses. However, on some campuses, a wide array of activities can be found, ranging from programs in international agricultural education to Peace Corps training programs to summer exchange programs. Agricultural educators are also occasionally involved in hosting foreign visitors--such as young farmers groups--who are interested in observing and meeting with teachers, extensionists, and farmers at the local community level.

Sponsors of on-campus teacher education activities in international agriculture are as varied as the activities themselves. However, the

U.N. Food and Agriculture Organization (FAO), the Ford Foundation, the Peace Corps, the U.N. Educational, Scientific, and Cultural Organization (UNESCO), the U.S. Agency for International Development (USAID), and the World Bank are organizations which continue to be active in sponsoring programs and personnel in international agriculture. Other sponsoring agencies include the governments of the participating countries and the host universities in the United States.

Involvement Abroad

If on-campus advising and teaching of foreign students is excluded as a criterion for "international involvement," then agricultural education faculties are more likely to be involved in international activities abroad than on their own campuses. Nearly one-third (31 percent) of the 84 agricultural education programs or departments participating in the AATEA survey reported involvement overseas during the period from 1960 to 1979. International activities overseas were most common, with those agricultural education programs located in the central states (56 percent). However, agricultural education departments in the eastern states were also active overseas, with over one-half (54 percent) of their programs involved abroad at some time during the 1960's and 1970's. Less than one-quarter (21 percent) of the programs in the southern states and only one (7 percent) of the departments in the western states reported institutional involvement abroad during the years studied. Departmental activity (number of institutions involved) abroad was somewhat greater (16 percent) during the 1970's than during the 1960's.

According to AATEA survey data, 26 agricultural education faculties were engaged in a total of 54 overseas projects in 36 countries during the past two decades. Nearly all those activities were conducted in the LDC's. Of these, 23 projects (43 percent) involved countries in Africa, 15 (28 percent) in Latin America, and 14 (26 percent) in Asia. Overseas projects were also conducted in two European countries. The countries involved are listed in Table 15.1.

Agricultural teacher education international projects typically involve one or more faculty members from a U.S. university, who are engaged in curricular, institutional, or pedagogical research and developmental activities at an overseas institution of higher education in agriculture. Those activities are usually conducted in coordination with the ministry of agriculture and/or the ministry of education of the host country. These projects are often under USAID sponsorship.

Institutions responding to the AATEA survey indicated that 80 different agricultural teacher educators were working abroad in departmental/program projects and activities at some time during the past two decades. Of this total, about one-half (39) of those individuals were still serving as active members of agricultural education faculties during the 1979-80 academic year. These figures included only those teacher educators who acquired overseas experience on university-sponsored projects. Despite this cadre of experienced professionals, most agricultural teacher education programs in the United States today still remain institutionally uninvolved overseas in international agricultural education.

TABLE 15.1--Countries involved with U.S. agricultural education departmental overseas projects by continent or area, 1960-1979

Africa	Latin America	Asia	Europe
(N = 23)	(N = 15)	(N = 14)	(N = 2)
Botswana	Brazil (4)	Bangladesh	Azores
Chad*	Columbia	Indonesia	England
The Gambia*	Costa Rica	Korea (2)	
Kenya (4)	Dominican Republic (2)	Malaysia	
Liberia	Jamaica (3)	Nepal (2)	
Mali*	Mexico (2)	Philippines	
Mauritania*	Panama	Saudi Arabia	
Nigeria (5)	Venezuela	Taiwan (3)	
Senegal*		Thailand	
Sierra Leone		Viet Nam	
Sudan			
Tanzania (2)			
Tunisia			
Uganda (4)			
Upper Volta* (2)			
Zambia			

Note: Parentheses contain the number of projects if more than one. Asterisk identifies each of six countries involved in one Sahel project.

Source: W. Thuemmel and R. Welton, AATEA Survey (15.1).

Curriculum and Program Offerings

Former Secretary of Agriculture Bob Bergland, in addressing the twentieth FAO conference in Rome in 1879, stated, "A principal U.S. goal is to help developing countries improve their own food production and distribution (15.9)."

Professional agriculturalists, educators, and scientists agree that agricultural and rural development problems must be solved through international cooperation. Solutions to those problems can be applied more and more in a worldwide context. To learn how universities with agricultural education faculties in the United States were programmatically attuned to the goal expoused by Secretary Bergland and other national leaders, Thuemmel and Welton asked respondents to the AATEA survey to identify the curricula, programs, and courses in international agricultural education which were offered by their institutions. Program titles identified at the undergraduate level were:

- Agricultural Economics Program with International Agriculture Emphasis
- Agriculture in Developing Countries
- Curriculum in International Agriculture (an option in the B.S. degree program in agricultural education)
- International Agricultural and Rural Development Study Opportunities
- International Agricultural Development (B.S. degree program)
- International Agricultural Education Option in Agricultural Science
- International Education
- International Studies Emphasis
- Semester Internship at a Foreign Institution of Higher Education
- World Food Issues

Graduate-level programs included:

- Agricultural Development (Master of Agriculture degree)
- International Agricultural and Rural Development (minor)
- International Agricultural Development (M.S. degree)
- International Agricultural Education (emphasis area in M.A., M.S., and Ph.D. degree programs)

Although most of the institutions included in the AATEA survey reported no program offerings in international agricultural education, several identified one or more internationally oriented courses in agriculture as being offered at their universities. Some examples of those course titles are listed as follows:

- Agriculture in Developing Countries
- Education for Rural Development
- Educational Programs in Agriculture for Developing Countries
- Extension Methods for Developing Countries
- Graduate Study in International Agriculture
- Independent Study in International Agriculture
- International Agriculture
- International Agriculture and World Food Problems
- International Agricultural Technology
- Seminar in International Agriculture
- Undergraduate Research in International Agriculture
- World Food and Population Problems
- World Food Economics

Agricultural educators also indicated that some of their traditional preservice and inservice courses have an international orientation. Although an international focus is not implied from their titles, courses such as "Methods and Materials in Adult and Extension Education" and "Planning Programs in Vocational Agriculture" can also be useful courses for students enrolled in international agriculture programs.

U.S. universities in general, and the agricultural education profession in particular, must be prepared to develop both undergraduate- and graduate-level programs and non-degree or certificate programs that will provide individuals with the competencies required for international agricultural projects. In a broader context, Smith has identified three major aspects of training in international agriculture with which colleges of agriculture are concerned (15.10). These are:

1. Improving the training of the increasing number of foreign students and professionals, both academic degree training and special training.

2. Increasing the international dimension in the training of all students in food and agriculture.

3. Providing specialized training for U.S. students and professionals who will be working in the international field.

Agricultural educators need to keep these aspects in mind when establishing program goals. It is also recommended that four factors should be considered in the development of all programs in international agricultural education. All programs should:

1. Include more in-country (home LDC) training for foreign students.

2. Focus on preparing participants to train their LDC counterparts.

3. Be practical in content and be relevant to the needs of LDC's.

4. Provide participants with the knowledges, skills, and attitudes needed to accomplish their task of facilitating agricultural progress in their own developing countries.

International Students

Many students from a wide variety of countries are enrolled each year in agricultural education programs at U.S. universities. According to data obtained in the AATEA survey, at least 245 students from 44 other nations were enrolled in agricultural education programs during 1978-79. Approximately two-thirds of those reported were enrolled in graduate programs. Nearly one-half (45 percent) of the foreign students were enrolled in southern universities, while at least an additional one-third (34 percent) were attending universities in the central states. A regional distribution of foreign student enrollments in agricultural education is presented in Table 15.2.

TABLE 15.2--Number of foreign students enrolled in agricultural education at U.S. universities, 1978-79

	Number by AATEA Region				
Level	Eastern	Central	Southern	Western	Total Enrolled
Undergraduate	5	17	59	4	85
Graduate	32	67	51	10	160
Total	37	84	110	14	245

Source: W. Thuemmel and R. Welton, AATEA Survey (15.1).

As would be expected, practically all the foreign students enrolled in agricultural education in the United States are from the developing nations. The identifiable countries with the largest numbers of agricultural education students in the United States in 1978-79 were as follows: Nigeria, 38; Iran, 29; Malaysia, 20; and Thailand, 12. A large majority (79 percent) of the total enrollees were from countries in Africa and Asia. A distribution of foreign agricultural education students by continent or area is shown in Table 15.3.

TABLE 15.3--Number of foreign students enrolled in agricultural education at U.S. universities by continent or area, 1978-79

Continent ( )[a] or Area	Number by AATEA Region								Total Enrolled
	Eastern		Central		Southern		Western		
	UG[b]	G[c]	UG[b]	G[c]	UG[b]	G[c]	UG[b]	G	
Africa (13)	4	11	13	29	7[d]	8[d]	0	1	73[e]
Asia (15)	0	8	2	28	18[d]	28[d]	2	4	90[e]
Europe (2)	1	3	0	0	0	0	0	0	4
Latin America (10)[f]	0	7	0	6	9	8	0	1	31
North America (1)	0	1	0	0	0	0	0	0	1
Oceania (3)	0	2	0	1	0	1	0	0	4
Origin not reported	0	0	2	3	25	6	2	4	42[e]
Total (44)	5	32	17	67	59	51	4	10	245[e]

[a]Number of countries represented.

[b]Undergraduate students.

[c]Graduate students.

[d]Countries were identified and numbers of students were reported, but the number of students by country was not specified by at least one respondent.

[e]Survey respondents reported 245 foreign students enrolled in their institutions (see Table 15.2), but only 193 could be identified by country of origin. Of those so unidentified, 10 were identified by continent of origin, 31 were thought to be from African or Asian countries, and 11 were untraceable.

[f]Includes the Caribbean region and Central and South America.

Source: W. Thuemmel and R. Welton, AATEA Survey (15.1).

Advisement

Wortman and Cummings have noted that:

> Many colleges and universities in the developed countries have high-quality undergraduate and graduate programs. However, while they may be strong in providing methodological skills, they are often weak in relating training to the problems of the country to which the student will return (15.11).

This caution is appropriate for agricultural education faculties in the United States, where foreign students tend to be admitted with the assumption that they will take the same curricula and courses and receive the same advisement services as are provided for U.S. students. It is very important for faculty advisors to foreign students to have a clear perspective of what their advisees will be expected to do when they return to home countries.

If agricultural education units have faculty members with international experience or interest, those individuals are usually assigned as academic advisors to the foreign students. However, many faculties do not have teacher educators with international experience or even interest. Since the needs and objectives of foreign students often differ from those of their host-country classmates, some agricultural education programs will need to strengthen the international focus of their staff and curricula if all students enrolled are to acquire an education appropriate to their professional goals.

Internship Requirements and Teacher Certification

The AATEA survey included questions on internship requirements (student teaching or similar field experience) for foreign students enrolled in agricultural education. The study also addressed the question of whether or not these students could be certified to teach vocational agriculture in the particular states where they were attending college. The majority of teacher education programs involved in the study indicated foreign students were not enrolled in their preservice programs. Most of the agricultural education programs with foreign students enrolled only at the graduate level during 1978-79 reported their foreign preservice program enrollees were not normally required to complete internship assignments. However, in sharp contrast, nearly all those departments with foreign preservice program students enrolled at the time of the survey required those students to complete some kind of internship assignment.

Wortman and Cummings have observed that:

> Agricultural research and extension workers in developing countries often are not willing to work at the farm level. Frequently this reluctance is based on the individuals' lack of technical and practical skills and lack of confidence in their ability to work with farmers; they fear embarrassment (15.12).

On a similar note, Smith has suggested that "practical experience and observations for foreign students should be increased since a majority of them have not had practical experience in agricultural production and in agricultural business (15.13)." These observations about foreign students also seem applicable today to many U.S. students who are enrolled in preservice agricultural education programs. With fewer and fewer students who have farm backgrounds enrolling in agricultural education it is becoming increasingly important that appropriate internship assignments be required of all agricultural education majors. This applies to domestic students as well as to foreign ones.

Several universities have established special procedures or arrangements to facilitate preservice internships for foreign students in agricultural education. Examples of the internship activities and/or procedures currently in use include:

1. Individualized internship activities adapted as to length and type of experience to the student's previous experience.

2. Additional observation and on-campus teaching.

3. Careful selection of cooperating teacher and teaching center.

4. Placement with a vocational agriculture teacher during the summer prior to student teaching.

5. Field work through observation and special projects. A student must complete equivalent hours in special projects in lieu of student teaching.

6. Cooperative extension work.

7. Participation in inservice activities.

Foreign students can be certified to teach vocational agriculture in slightly over one-half the states. However, most foreign students in agricultural education do not pursue teacher certification in the United States, even in those states where such is permitted.

Teacher Educator Experience and Interest

The key to institutional involvement, program development, and student recruitment with regard to teacher education activities in international agriculture lies within the agricultural education profession and with the various universities where those professionals are employed. However, only a relatively small proportion of U.S. agricultural teacher educators have overseas experience in professional agricultural activities. Although Trotter listed 348 teacher educators in the 1978-79 Directory of Agricultural Teacher Educators (15.14), it was previously noted that as few as 39 of this total (about 1 in 9) had acquired international experience abroad in projects involving their agricultural education departments or programs and were still serving on those faculties during the 1979-80 academic year. Results of the AATEA survey also showed that an additional 33 agricultural teacher educators were experienced in international agriculture. Unlike the first group, however, these educators acquired their overseas experience in projects which had not directly involved the agricultural education departments where they were employed at the time of the survey. Thus, a total of approximately 72 agricultural teacher educators (or about 1 in 5 of those serving on agricultural education faculties in 1979-80) had overseas experience in international agricultural education.

Many agricultural teacher educators without overseas experience have expressed interest in becoming involved in international activities. In fact, in the southern and western states, this group was larger than those with experience. In response to a question on the AATEA survey, 140 of the teacher/extension educators (40 percent) indicated they had experience and/or interest in international activities. This figure included those "with experience," who were assumed to also "be interested." The percentage of teacher educators with experience and/or interest in international activities dropped sharply and uniformly in the regional pattern as follows: eastern, 62.9 percent; central, 49.5 percent; southern, 33.5 percent; and western, 15.2 percent. A percentage distribution of teacher educator international experience and/or interest is presented in Table 15.4. The data show that two-fifths of the agricultural teacher educators in the United States are experienced or interested in international activities.

TABLE 15.4--Percentage distribution of teacher/extension educators with experience and/or interest in international activities in agriculture by U.S. region, 1978-79

Experience and/or Interest	Percentage by Region				
	Eastern	Central	Southern	Western	Total
	(N = 54)	(N = 99)	(N = 149)	(N = 46)	(N = 348)
Experienced	40.7	27.3	15.4	4.3	21.3
Interested	22.2	22.2	18.1	10.9	19.0
Not interested	11.1	17.2	36.2	8.7	23.3
No response	25.9	33.3	30.2	76.1	36.5

Source: W. Thuemmel and R. Welton, AATEA Survey (15.1).

Countries Where International Experience Was Acquired

Data obtained from agricultural teacher educators who completed overseas assignments prior to 1979 suggest a broad range of experience in international agriculture. Seventy-two teacher educators indicated foreign service in 67 countries. Although international assignments were concentrated more heavily in the western hemisphere, teacher educators reported experience on every continent except Antarctica. Fifteen indicated foreign service in Brazil, nine in Nigeria, nine in Venezuela, seven in Kenya, six in Jamaica, five in India, five in the Philippines, and five in Mexico. Less than five teacher educators were believed to have served in each of the other countries reported. The countries in which agricultural teacher educators acquired experience in international agriculture prior to 1979 are identified in Table 15.5.

Nature of Assignments

Agricultural teacher educators with overseas experience were asked to identify the type of service associated with their international assignments, in other words, the nature of each assignment. By using a classification system similar to one used in an NACTA study by Rawlings and Foutch (15.15), Thuemmel and Welton were able to summarize the varied responses of these teacher educators. The following categories were used: (1) consulting, (2) research and development, (3) teaching, (4) administrative, (5) extension, and (6) other. Some respondents indicted only one type of service, while others reported several. The extent of service in each category is shown in Table 15.6.

Employing Agencies

Teacher educators were also asked to identify their international assignment employers. At least 47 different agencies were listed by the respondents. The U.S. government--specifically the USAID--was the employer identified most frequently. Other employer categories or agencies frequently cited were private firms and/or foundations, universities, foreign governments, and international organizations. The Ford Foundation was the largest employer among the private firms and/or foundations listed. Although several universities (both U.S. and foreign) and university consortia were identified as employers, it was believed that in some cases the employing universities were actually administering contracts using funds provided by the USAID or by national governments. The National Institute for Cooperative Education of Venezuela accounted for over one-half of the employment provided agricultural teacher educators by foreign governments. Teacher educators also worked overseas for international and religious organizations.

Using the Rawlings and Foutch classification system, Thuemmel and Welton summarized the types of employing agencies and the number of teacher educators employed. Their results are indicated in Table 15.7.

TABLE 15.5--Countries by continent or area in which active members of U.S. agricultural education faculties in 1979-80 reported completing professional assignments between 1960 and 1979

Africa	Asia	Latin America	Europe
(N = 23)	(N = 16)	(N = 15)	(N = 7)
Benin	Afghanistan	Brazil (15)	Azores
Botswana	Cyprus	Colombia (3)	Denmark
Chad	India (5)	Costa Rica (4)	England (2)
Egypt	Indonesia	Dominican Republic (3)	France
Ethiopia (2)	Iran (4)	El Salvador	West Germany
The Gambia	Japan (3)	Guatemala	Italy (2)
Ghana	South Korea (3)	Guyana (2)	Portugal (2)
Kenya (7)	Lebanon	Jamaica (6)	
Lesotho	Malaysia (2)	Lesser Antilles	
Liberia (2)	Nepal (2)	Mexico (5)	
Malawi	Pakistan	Nicaragua	
Mali	Philippines (5)	Panama	
Mauritania	Sri Lanka	Paraguay	
Nigeria (10)	Taiwan (3)	Peru	
Senegal	Thailand (4)	Venezuela (9)	
Sierra Leone (2)	Turkey		
Somalia			
Sudan (2)			
Tanzania (4)			
Tunisia			
Uganda (3)			
Upper Volta (2)			
Zaire			

Oceania
(N = 7)
American Samoa
Australia (2)
Melanesia (2)
Micronesia
New Zealand (3)
Papua New Guinea
Western Samoa

Note: Parentheses contain the number of agricultural teacher educators (if more than one) with experience in those countries.

Source: W. Thuemmel and R. Welton, AATEA Survey (15.1).

TABLE 15.6--Nature of international assignments experienced by agricultural educators

Type of Service	Number of Teacher Educator Assignments
Consulting	68
Research and development	22
Teaching	17
Administrative	16
Extension	2
Other[a]	7
Total	132

[a]Includes conference participation, farming, sabbatical visit, trade mission representative, and work-study assignments.

Source: W. Thuemmel and R. Welton, AATEA Survey (15.1).

Language Proficiency

Approximately one-third (47) of the teacher educators with experience or interest in international activities indicated some knowledge of a language other than English. Respondents reported various degrees of proficiency in 11 foreign languages. Fifteen of those individuals claimed some skill in more than one foreign language. The major proficiencies were the ability to understand, speak, read, and write Spanish, French, Portuguese, and German. Proficiencies tended to be highest in reading and lowest in writing. A detailed summary of foreign language proficiency is presented in Table 15.8.

Interest in Future Assignments

Most (86 percent) of the respondents with experience and/or interest in international activities also reported interest in future assignments abroad in agricultural education. The same group of teacher educators checked "consulting" most frequently as the kind of foreign service which they would prefer. "Teaching" was ranked second and "research and development" third. A summary of the kinds of foreign service preferred is shown in Table 15.9.

TABLE 15.7--Agencies that employed agricultural educators for foreign service prior to 1979

Agency	Number of Teacher Educator Assignments Supported
U.S. government	
Agency for International Development	44
Peace Corps	8
Fulbright Program	2
Other	5
Private firms and foundations	21
Universities	18
Foreign governments	12
International organizations	12
Religious organizations	4
No agency identified	5
Total	131

Source: W. Thuemmel and R. Welton, AATEA Survey (15.1).

Teacher educators indicated much variety in their first-, second-, and third-choice locational preferences for overseas work. As requested, some responses identified specific countries, while others included general areas such as regions or continents. Countries mentioned included Australia, Brazil, China, Colombia, India, Jamaica, Nigeria, Panama, Portugal, Tanzania, and Tunisia. When specific countries were cited, they were usually included among the choices of individuals who had completed previous assignments in those nations. No one country was highly preferred over the others. Latin America (Central and South America and the Caribbean area) and Europe were the most popular area preferences. Country, regional, and continental choices were consolidated into continental and area preferences and are presented in Table 15.10.

Teacher educators exhibited a wide range of flexibility in their preferred length of overseas assignment. Thirteen of the 121 respondents did not indicate a preferred length of assignment and, of those who did, most indicated more than one length of assignment preference. Their responses were as follows: short-term (3 months or less), 88; intermediate-term (4-12 months), 25; and long-term (over 12 months), 17. In summary, although a majority (72 percent) favored short-term assignments, most teacher educators would look favorably on longer assignments as well.

TABLE 15.8--Foreign language proficiency, as reported by agricultural teacher educators, 1979 (N = 140)

Language	Numbers by Type of Proficiency			
	Understand	Speak	Read	Write
Spanish	20	20	26	15
French	6	2	12	1
Portuguese	8	7	8	3
German	5	4	5	4
Farsi	2	2	1	1
Swahili	1	1	1	0
Czech	1	1	0	0
Italian	0	0	1	0
Arabic	0	1	0	0
Nepalese	0	1	0	0
Tagalog	0	1	0	0

Source: W. Thuemmel and R. Welton, AATEA Survey (15.1).

Some Considerations for International Assignments

Most of the chapter thus far has presented an overview of institutional and teacher educator international activities during the past two decades. Some implications have also been drawn in assessing the agricultural education profession's potential for contributing to international agricultural and rural development. The remaining pages of the chapter will focus on what factors students and agricultural educators need to consider about international assignments and on how they can gain access to opportunities in international agriculture.

Persons contemplating entry employment into an international agricultural program should first make a self-appraisal of their attitudes toward international work. The following checklist of questions has been designed by McCreight to help individuals in identifying personal strengths and weaknesses as related to entry into international employment (15.16). Affirmative or negative responses are required for each question.

TABLE 15.9--Kinds of foreign service preferred by agricultural teacher educators

Kind of Service	Number of Preferences
	(N = 121)[a]
Consulting	95
Teaching	77
Research and development	53
Administration	35
Extension	31
No special preference	13
Other	7
Total	311[b]

[a]Nineteen of the 140 agricultural teacher educators with experience and/or interest in international activities did not respond to this item.

[b]The total exceeds the number of respondents because more than one choice was requested per individual.

Source: W. Thuemmel and R. Welton, AATEA Survey (15.1).

ATTITUDE QUESTIONNAIRE FOR

INTERNATIONAL EMPLOYMENT

+	−	
____	____	1. Do you and/or would you like traveling to foreign countries?
____	____	2. Would you like to live and work in developing countries?
____	____	3. Would you like to work with people classified as rural poor?
____	____	4. Are you interested in speaking, reading, and writing other languages?
____	____	5. Do you believe that most agricultural and extension education programs in the United States can be directly transferred and implemented in developing countries?

____ ____ 6. Do you think socio-cultural aspects should be considered in designing and implementing programs in developing countries?

____ ____ 7. Would you accept working through lengthy bureaucratic processes to obtain approval for actions?

____ ____ 8. Would you be impatient if supplies and equipment arrived several weeks late?

____ ____ 9. If things weren't progressing well, could you make changes to reach your objective?

____ ____ 10. Would your spouse and/or family object to your being away from home for periods of up to two months?

TABLE 15.10--Continents or areas preferred by agricultural teacher educators for overseas assignments (N = 140)

Continent or Area	Frequency of Choice[a]		
	First	Second	Third
Latin America (including Caribbean area)	34	31	20
Europe	33	15	10
Africa	19	22	10
Asia	9	11	16
Oceania	3	11	7

[a]No preference was indicated by 23 respondents; 19 did not respond to this item.

Source: W. Thuemmel and R. Welton, AATEA Survey (15.1).

An interest in traveling and living in foreign countries and working with rural poor people is essential to the development of a positive work attitude toward international projects. Therefore, an affirmative answer to the first three questions indicates a positive interest toward future international employment.

Although many counterparts in international work may speak English, it is essential that an attempt to speak, read, and write the native language of the developing country be made. Again, a positive response to Question 4 is another positive indication for international assignments.

After one has experienced international work, it will be obvious why Question 5 should have been answered negatively. In developing countries, the simple transfer of existing U.S. programs cannot occur with the expectation of the same or equal results. Modifications must be made to account for socio-cultural aspects and the current status of development within the country. An affirmative response to Question 6, considering the socio-cultural aspects for project design, is another positive response.

The bureaucratic process involved to acquire approval for actions is a lengthy one in most assistance agencies. Possession of a high degree of tolerance for time-consuming processes that usually involve a number of foreign and domestic governmental agencies is necessary. A negative response to Question 7 reveals a needed understanding for the necessity of a high degree of tolerance.

Acquisition of supplies and equipment is very slow in developing countries. Add transportation difficulties to the bureaucratic process, and it can be understood why a negative response should have been given to Question 8.

Flexibility, adaptability, and innovativeness are essential traits for successful international employment. It is very important for international consultants and administrators to make the necessary changes to achieve project objectives.

Last, but of prime significance, is an understanding spouse and/or family. International projects often require personnel to be separated from their family members for periods of two or more months. With strong objections to such absences, international employment could be extremely difficult.

Therefore, positive responses should have been given for Questions 1, 2, 3, 4, 6, and 9 and negative responses given for Questions 5, 7, 8, and 10 for the highest attitude rating when international employment is being considered.

The following rating scale is designed to assist in the determination as to whether international employment should or should not be considered, based on the Attitude Questionnaire for International Employment.

INTERNATIONAL EMPLOYMENT ATTITUDE

RATING SCALE

9-10 correct responses	Excellent
8 correct responses	Good possibility, but some guidance is required
7 correct responses	Acceptable, but considerable guidance is required
6 correct responses	Questionable
5 or fewer correct responses	Should not consider international employment

In summary, to succeed in international employment, one should:

- Be interested in traveling and in the type of work.
- Have empathy for rural poor people and for people of different cultures and different languages.
- Be flexible.
- Be adaptable.
- Possess a high degree of tolerance.
- Have an understanding spouse and/or family.

Gaining Access to Opportunities in International Agricultural Education

Preparation for and access to international employment are complex and multi-dimensional tasks. The following are suggestions for anyone beginning this process.

- Develop some expertise in a technological field which is internationally marketable (agriculture is an excellent choice).
- Acquire proficiency in those behavioral sciences which are related to the learning, communicative, and human development processes.
- Have some understanding of and appreciation for administrative (bureaucratic) procedures.
- Be able to apply all the previous factors to a complex socio-cultural setting (which probably has not yet been selected).

Fortunately, there are many avenues of approach to acquiring these qualifications. How one gains access to international opportunities will be influenced by his/her current level of professional development.

Suggestions for Professional Agricultural Educators

Few formal programs exist to prepare a professional for international employment; consequently, it is usually necessary for him/her to plan and prepare on an informal basis. Assuming a person has had formal training in a specialized area needed for agricultural assistance projects, the following preparation would be desirable for entry into international project work as a consultant.

1. Learn to speak, read, and write at least one useful language. Currently, French and Spanish are most in demand for qualifying to work in the developing countries.

2. During a sabbatical or another type of leave from employment, travel to developing countries to observe on-going programs and, if possible, to serve a brief internship with an international project. Also take extended vacations outside the United States, preferably in developing countries.

3. Attend international meetings. Involvement as an instructor and as a participant at international seminars with individuals from developing countries can further one's understanding of the international setting and its problems as well as enhance one's professional development.

4. Take part in on-campus international activities and programs. This can heighten one's understanding of cross-cultural similarities and differences and, at the same time, lead to information about overseas employment opportunities. One should maintain close contact with the university's international student advisory office, work with the campus Peace Corps coordinator, advise foreign students, and maintain communications with the individual responsible for international programs in the college of agriculture.

Advice for Undergraduate Students

Teacher educators experienced in international work offer the following advice to agricultural education students interested in pursuing some phase of international employment.

1. Select a university that supports and is actively involved in international programs. During undergraduate enrollment, participate in international activities on campus and regularly contact international program officers, for these activities can provide initial experiences toward understanding the international setting.

2. Pursue a major area of academic concentration which will allow one to become identified as a specialist in a needed field of expertise. Be sure to acquire as much practical experience as possible in the area of specialization. To qualify as a specialist, one must obtain graduate-level training.

3. Learn a useful second language. Also, consider enrollment in anthropology and sociology courses to enhance understanding of other cultures. Get acquainted with international student classmates. Learning another language and understanding other cultures are very important in preparing for international work.

4. After obtaining vocational teacher certification, get three to five years of teaching experience, begin work on an advanced degree in agricultural or extension education, and consider volunteering for Peace Corps duty.

5. When choosing a graduate program advisor, select an agricultural teacher educator with international experience.

Summary

During the past two decades, the dynamics of an increasingly interdependent world have heightened the international need for development assistance specialists--especially agricultural technicians and educators. Most U.S. universities have been and continue to be involved with international activities; however, this has not been the case with many agricultural education faculties. Only one-half the agricultural teacher education programs in the nation indicated involvement in international activities during the 1960's and 1970's. Nearly all those activities were conducted in the developing countries.

Title XII of the International Development and Food Assistance Act of 1975 has provided a mechanism for agricultural educators to become institutionally more involved in international agricultural and rural development projects during the 1980's. Agricultural teacher educators who have overseas experience constitute an important nucleus for the growth and development of international agricultural education in the United States. Not every teacher educator is suited for, nor desires, international experience; however, many overseas opportunities are available to those agricultural educators who seek them out. International experiences can be personally and professionally very rewarding.

In the past, the role of teacher education in international agriculture has been significant, but limited. This indicates a relatively low priority within the total scope of professional activities. It is recommended that a greater number of agricultural teacher educators and their departments become involved in international activities in the years ahead. The agricultural education profession has much to contribute to overseas development. Greater international concern and involvement will have an invigorating effect upon agricultural education, both in the United States and in other countries of the world.

CHAPTER 16

Philosophy for Teacher Education in Agriculture

Gordon I. Swanson
UNIVERSITY OF MINNESOTA

This chapter has a threefold purpose: to eliminate some of the vagueness and confusion which exists about an important aspect of individual and institution life--namely, philosophy and its close relative--history; (2) to provide a framework for understanding philosophy and history; and (3) to offer some perspectives about the importance of philosophy to individuals, to institutions, and to society.

While these are achievable, the purposes served by philosophy itself remain unfinished. None of the problems of human existence is more persistent or more fundamental than the problem of means and ends, which involves ensuring that our methods, competencies, and techniques are adequate to the ends we seek; that the ends we seek are reasonable and relevant; and, most important, that the ends are not determined, dominated, or supplanted by our chosen means. Constant vigilance to ensure that the things which matter least are not traded for the things which matter most is a part of the role of philosophy. Thus, its status is always that of unfinished business.

What Is Philosophy?

There is not widespread agreement on a definition of philosophy; but, there is general agreement on the observation that it can be viewed as an academic discipline, a subject that can be studied in schools and colleges. As an academic discipline, it is concerned with logical behavior--the nature of reality and the nature of knowledge. Through the use of powerful tools of inquiry, mainly those of synthesis and deduction, the subject is pursued with vigor and with dedication.

As an academic discipline, philosophy is well known for having spawned a number of schools of thought. Some of these have won advocates and have assumed the status of separately identifiable philosophies, such as positivism, idealism, reconstructionism, and existentialism.

Ethics, a subset of the discipline of philosophy, was once studied entirely within the domain of discipline-oriented philosophy. It has since moved elsewhere, mainly to professional schools, where its practical considerations are paramount.

A second view of philosophy is that of a set of beliefs, or values, held by individuals to guide their behavior or that of their institutions. It is the view often taken by employers who ask prospective employees to "write a statement of your philosophy." Similarly, professors often ask students to write a statement of their philosophy. In such instances, it is assumed that one's set of beliefs and values have certain qualitative characteristics of rightness or wrongness which can be judged on a scorecard. It is also assumed that everyone has such a philosophy and that there may be a certain catharsis in writing it. What other value could there be? Neither the prospective employers nor the professors are anxious to give the "correct" answers, which would yield a higher score on the scorecard.

The beliefs-and-values view of philosophy is the most widespread and thus the most commonly held view of philosophy for many fields, including agricultural education. At the same time, it is also the most easily confused with ideology, a construct which is often inimical to philosophy. An ideology operates under conditions in which a belief or value is held no matter what! Once a belief enters the realm of ideology, truth need not be further pursued; it has already been found! Under such conditions, the means-ends problem diminishes and the ideologically framed philosophy is no longer unfinished business. While the beliefs-and-values expression of philosophy is vulnerable to expressions of ideology, it need not occur. The test is whether beliefs and values remain open to change and whether the conditions for change are available.

The danger of ideology need not be in terms of examples which are obscure, such as Marxism, or remote, such as some distant cult. Comfortable detachment from distant ideology is a luxury. Comfortable attachment to proximate ideology is a danger. What is the status, for example, of the FFA? Is it a means or an end? Is it a tool in the process of being examined, improved, or even discarded, if necessary, in order to replace it with a tool more adequate to the ends being sought? Does the organization and its leadership spend more time polishing its image or more time defining, clarifying, and seeking the goal of becoming a better tool? Has the truth which was sought been found, or is the search still underway?

How these questions are answered will determine whether this youth organization is a part of the field's philosophical unfinished business or whether it is some comfortably attached ideological baggage. Many similar examples can be found close at hand.

There is a third view of philosophy, a view which regards it as a set of standards anchored in propositions or premises which are constantly subject to inquiry and to re-examination. It is this view which allows graduate schools to grant the Doctor of Philosophy degree to individuals who have never taken a course in philosophy. It is the existence of standards and the interest in applying them which are important, not the subject matter to which they are applied. If philosophy is not something which drifts with a light anchor and if its anchor points are identified with inquiry-related standards, how is the testing done? By empirical verification and by tests to determine consistency with reality, and by an unending search for verification and for consistency.

Empirical Verification

If a standards view of philosophy requires the rigor of empirical verification, what must be tested or verified? Again, it is important to recall that philosophy is a problem of defining ends and means and to reiterate that it is ends which demand verification. But, empirical verification does not consist of finding a rationalization or a defense of the ends pursued. If the ends of agricultural education include the strengthening of communities, the widening of individual choices, and the exercise of freedoms, these ends can be examined and subjected to rigorous empirical tests.

While focusing on ends, one should also attempt to empirically verify those means which have come to dominate education and to serve as proxy ends. Education's preoccupation, for example, with cognition and with conceptual-verbal abstractions may be an eligible target. As certain means begin to dominate as surrogate ends, they may crowd out the access routes to ends such as the esthetic aspects of life or the values derived from work, values which are verified most easily when individuals are deprived of esthetic values or prevented from working.

Empirical verification of the basis for choosing ends for education cannot, however, be approached as if it were invariably a clean, data-centered problem. The ends of education are intended to create an advantage, and advantages are rarely distributed or accepted equally. The problem is not without its share of politics including, occasionally, some academic imperialism. There is also need, therefore, for tests of reality, including the reality of political power.

Consistency with Reality

If a standards-oriented philosophy involves anchor points which are consistent with reality, what is the test of reality? It is history; all of reality is history. It is the reason why history, particularly educational history, is a close relative of the philosophy of education.

While all reality is history, all recorded history may not be a careful interpretation of reality. Indeed, the writing of history is always purposeful, and its purposes can also be to obscure reality or to create perceptions about a reality which never occurred. It is a misuse of history, therefore, merely to assemble the writings of the past and to accept the assertions of the writer at face value. Critical judgment should be applied to all observations, whether made by experimental reseachers or by historians.

In this pursuit of history-based reality, it is useful, to take a side trip to discover some of the uses of history and some of the ways of viewing it. History may be seen as:

1. A flow of forces affecting our values, our way of thinking about what is important. For example, these forces influence our way of thinking about the reformation, civil rights, military conscription, and freedom.

2. A way of looking at the nature of change, including the propensity to destroy in order to create as is often seen in the rhetoric of political conventions and campaigns. The historical nature of change can be seen in many interesting ways. Changes do not always occur as a smooth flow of inertial forces. Occasionally, they occur through crises specifically created to induce change.

3. A way of viewing some drastic change in the direction or flow of events. For example, such a drastic change occurred with the war of independence, the creation of the land-grant university system, with the civil rights movement, and with many similarly drastic changes. Each of these becomes a favorite topic for historical stock-taking.

4. A way of looking at the arguments or debates of the past which continue to influence the decisions of the present. The best example is the nation's entire system of jurisprudence. It is the debates of the past, won or lost, which add to the arguments of the present.

5. A way of looking at trends in representative governance or, more specifically, in legislation. A historical look at legislation shows, for example, that most educational legislation is retrospective--it is fashioned to solve problems perceived a decade or more in the past. The reason, simply, is that public opinion responds to stimuli; it does not anticipate events. Accordingly, a legislative response to public opinion is a response to needs perceived in retrospect.

   And there are many other ways of illustrating the ways in which history can reveal interesting trends in legislation. Vocational education legislation has often been triggered during or immediately following periods of war, recession, or threats to national security. Its other trends involve slow and incremental changes, often grudgingly accepted, toward democratizing the system in which it functions.

In addition to these various ways of viewing history, there are also ways in which history may be used or even exploited, although not necessarily with any intention to deceive. Two illustrations of such uses are:

1. A way of creating selective perceptions of reality. Since all expressions of history are selective, any written history is biased by the nature of the selectivity of the writer. There is no complete record of either the present or the past, and memory, like preference, is highly selective.

   The writer of history becomes, therefore, a wielder of power, power in the selective perception of events and consequences. Since the winner remains to write the history of the battle, it should never be a surprise to observe the battle being described as good prevailing over evil.

   History is rarely seen objectively. It is often seen through the lenses of ideologies, biases, or prior loyalties such as socialism, liberalism, political party preference, or agency orientation. This is the obvious reason why it is possible to read historical accounts of the development of agricultural education written under the aegis of the U.S. Department of Agriculture that make no reference to the existence of vocational agriculture. Similarly, it is possible to read historical accounts of the development of programs of worker training written under the aegis of the U.S. Department of Labor that have little or no cognizance of the existence of vocational education.

2. A way of exploring the history of the various uses of educational history. An example involves the history of education in the United States. Earlier historical accounts dealt almost exclusively with a record of how education was institutionalized, a record of the origin of schools, together with a record of how they were organized and governed. A second wave of the history of education dealt with the influence of education, with schools included as merely one of the many influences. A more recent emphasis involves the record of educational imbalance or inequality, the record of why certain groups of citizens have received much more, or less, education than others.

The preceding which offered some differing views of history as well as its uses and abuses should be sufficient to show the unfinished and incomplete nature of the contributions of recorded history to contemporary thinking. It should also illustrate the need for critical judgment of historical evidence if one is to be concerned with rigorous standards for truth-seeking, with relating ends to means, and with ensuring that history, particularly recorded history, is an accurate picture of reality.

It also is useful to observe that there are no historians of vocational agriculture, much less of its essential corollary--teacher education. While it is possible to find a record of events, legal authorizations and institutional decisions effecting or affecting agricultural education, it is not possible to find individuals who invest significant amounts of time scouring original sources and who synthesize the past in search of hypotheses for instructing the present or the future. The contribution of immigrants to agricultural education or the accommodation of immigrants to a monolingual language of instruction is among the stories which are unresearched and unwritten. The ignorance of history remains, therefore, as the mother of innovation. Success mechanisms of the past are often discarded, and mistakes are repeated.

This discourse also serves as a preface to a brief discussion of the most historically relevant, as well as the most contemporarily vivid, of all the forces which have shaped the structure of philosophical thinking, namely, the influence of the roots of culture. Often referred to as cultural barriers, confrontations, or loyalties, they were first noted by Herodotus, the father of history, when he began to chronicle cultural conflict. Seen historically or as current events, these include, for example, cleavages between the Persians and the Greeks, between Christianity and Islam, between the have's of the northern hemisphere and the have-not's of the southern, and conflicts between Marxism and capitalism.

In addition to providing a framework for values, such conflict provided the basis for many important lessons. For example, cultural conflict often produces many victims--the displaced persons who live in poverty because they happen to have had the wrong allegiance at the wrong time and in the wrong place. Other victims include those who have no knowledge of the numbing effect of cultural conflict on human intellect.

Cultural conflict has lessons to teach about arrogance, not the arrogance which is born of confidence with one's own condition, but rather from the assumption of superiority of a chosen loyalty whether it be an ideology, a community, or a race. The great conflicts are not between those who identify themselves as human but between the organizations and the loyalties with which humans identify. A large share of cultural conflict is rooted, for example, in religion.

The existence of cultural conflict teaches how cultures allow the patterns of the past to guide visions of the future. All cultures have elaborate mechanisms for preserving and reconstructing the past. Occasionally, these mechanisms have reconstructed a past which is more appealing than the one which actually existed.

The more interesting lessons occur where there is cultural pluralism, where multiple streams of cultural influence join or converge to form one stream, but where the constituent streams are sufficiently incompatible so as to retain their original, value-oriented identity. This is the situation in the United States. It is a situation which demands a clear view of the way in which the streams of cultural influence tend to affect the philosophic behavior of individuals. To illustrate this phenomenon, we will discuss three constituent streams--the philosophical, the theistic, and the socio-economic.

The Philosophical Stream

The philosophical stream originated in ancient Greece. It may be unwarranted to refer to it here as the philosophical stream, since all subsequent streams will be described as having a philosophical influence. Yet, this one of Greek origin is so fundamental to the concept of philosophy that it will be distinguished by the generic label. After all, the word "philosophia" is of Greek origin, meaning "the love or pursuit of wisdom," a clear designation of purpose.

Well-known contributors to the influence of this stream are individuals such as Socrates, Plato, Aristotle, and their careful biographer, Plutarch. The influence of this stream stems from its powerful concepts, its unswerving commitment to standards, and its unfinished pursuit of truth. Its powerful concepts include those of democracy, freedom, and liberty. In early, yet fundamental, forms, these concepts originated and continued as a part of the stream. They have become a part of the fabric of the United States.

This stream's commitment to standards and to the expression thereof has become so completely integrated into western culture that their origin has often been obscured. The standard for architecture in the model of the Acropolis and the Parthenon has been followed in the construction of many government buildings, banks, and libraries. The standard format for musical composition, as provided in Aristotle's Da Musica, became the accepted standard for more than 20 centuries.* The accepted format for drama, three-act plays, has become so completely accepted that one hears only the references to deviations from the standard, for example, one-act plays. Much that is standard related originated within this stream.

---

*The Unfinished Symphony by Shubert was, in fact, finished. But it did not measure up to the Aristotelean standard. To demonstrate that the composer was not unaware of the accepted standard, he merely called it "unfinished."

There are numerous contemporary examples of the extent to which existing western culture strives to associate itself with elements of prestige, status, or standards associated with this stream. Prestigious organizations or associations have, for example, adopted the designation of academy, the name of the art school created by Plato. All secondary schools in France are still called lycées, a term taken directly from the word "lyceum," the name of the school created by Aristotle. An unquestioned subject in the secondary schools is the subject which was accorded a lofty position in early Greek culture--physical education.

Yet, it is the unending pursuit of truth which wins this stream its designation as a philosophical stream and which describes philosophy, as in the early paragraphs of this chapter, as unfinished business. Plato's Academy offered no claims that it was teaching the truth; it only claimed to be pursuing it. The pursuit was the end, and any discovery of truth was regarded as transitory, temporary, or instrumental and thus only a means to illuminate a further search.

This commitment to the pursuit of truth may have had an enormous influence on the early Christian church. The major literature of Christianity, the New Testament, was written in Greek. Indeed, much of it was addressed to the citizens of Corinth, Thessalonika, and Philippi. It should have been quite natural, indeed almost pro forma, for the church to have established its headquarters or its base of operations in Greece. In Greece, there was almost no record of individuals being persecuted for their beliefs; an ideal place, one would think, for the newly developing Christian church.

However, the church did not hesitate to reject the idea of a home base in Greece. All hardships paled alongside the one which is always troublesome--how to deal with truth. The Greek influence urged a continued search, and the early church was certain that it already possessed it. The church preferred to be where its possessions would be more honored than questioned.

This philosophical stream, while providing strong roots, was only partially compatible with the evolving American culture. Democracy in Greece was not available to all citizens. It was not available to women nor to slaves, the major working classes. Those for whom destiny or duty required physical work or attention to the economic relationship of the community were not accorded the status nor the rewards given to others. Indeed, freedom came to be viewed as being liberated from the demands of work in order to pursue what was simultaneously interpreted as a higher goal, a dual interpretation which has persistently accompanied this stream in its influence on American culture.

The Theistic Stream

The theistic stream had its origin in the birthplace of the Judeo-Christian tradition. Its influence on American culture is so powerful that it needs very little re-emphasis here. From the sixteenth century onward, it became the central force for the organization of much of the voluntary behavior at the village or community level, and it became the central feature for expressions of the highest level of the state of knowledge in music, sculpture, architecture, art, and craft work. It became a dominant force in governance as well as in education.

This stream contributed two primary influences to the mainstream of American culture: an orientation to work and a focus on the individual.

The orientation to work was not originally a part of this stream. In earlier years, work was regarded as punishment for sins and an avenue for its atonement. From the sixteenth century onward, work, within this stream, began to acquire some intrinsic merit, some value not necessarily identified with prior transgressions. Indeed, it was often regarded as having some of its own soul-redeeming features. It became known as the Protestant work ethic, a term which was an encomium, not a description; it was neither mainly Protestant nor necessarily ethical.

The other main influence, the focus on the individual, has numerous expressions in contemporary American culture. It is seen in the emphasis on individualized instruction, counseling, career planning, and goal setting. Its roots are deep in the theistic stream where salvation, for example, has always been an individual matter. Its contrast may be seen in Marxian theology, wherein individual decisions are subsidiary to the decisions of the group and where individual freedom in microcosm, on a grand scale, is regarded as the cult of the personality.

The Socio-economic Stream

The socio-economic stream which flowed into American culture originated mainly in England. Its impetus was strengthened at a historic time, in 1776, by a strategic volume entitled The Wealth of Nations, written by Adam Smith. It was an explanation of why some nations were more favored than others, and it became a classic of economic literature. It became convenient thereafter for the new nation, the United States, to be influenced by it.

The volume was not revolutionary; it merely reflected upon and interpreted the conditions in England. The social conditions in England included a rigid caste system, rigorously defended elitism, and various types of segregation-- particularly segregation of the poor. What originated as classical economic literature became, therefore, classical social theory. It reaffirmed the conventions of English culture, and further, it ascribed these conventions to God's will.

Smith's work provided justification, not reasons, for the wide differences which occurred in British society--the workers, the peasants, and the poor at one end of the scale and the industrial leaders, the professionals, and the politicians on the other. He justified such differences as being expressions of natural law, a system of harmonizing the differences which existed among individuals in order to promote the general good through divine sanction. He urged that people be taught to accept their stations in life and not to aspire to higher rungs on the occupational ladder. Since all of this was governed by the "invisible hand," the plight of peasants was asserted to be as divine as the right of kings.

This was the socio-economic stream which flowed into American culture. It was class oriented; it did not encourage respected institutions to train people for work, and it devalued the role of laborers and farmers. The concept of a gentleman was that of a person with undirty hands who regarded workers as management problem for the elite and never as a development opportunity for either the workers themselves or society.

One cannot deny the import of this stream on U.S. history nor on the contemporary scene. Poor farms were established in nineteenth century America as a counterpart to the English Poor Laws. Educational institutions were established along hierarchical lines to serve the hierarchical patterns which this stream endorsed. The formation of free trade unions and farmers' organizations came into existence as rebellions against the influence of this stream.

The Confluence of Streams

Earlier it was stated that cultural pluralism in the United States has historical derivatives and that the derivatives exist as streams of influence forming a mainstream but remaining therein as essentially incompatible. Three have been elaborated, and there are undoubtedly more existing as tributaries originating in Africa and Asia. All contribute to the reality of which history informs and to the often incompatible values of the theistic stream conflict, for example, with the class-conscious values of the socio-economic stream. The concept of freedom introduced by the philosophical stream is not the freedom which is reflected in other streams. It is now possible to postulate that all or most of our social problems--affirmative action, civil rights, equality of opportunity, youth unemployment, bilingual education, and many more can be traced to the incompatibility of the value orientations imbedded in the separate streams which make up our unique form of cultural pluralism. Only an understanding of history can illuminate this pluralism.

History Specific to Agricultural Education

The tracing of the cultural streams which accounts for a special kind of value-differentiated cultural pluralism is a very general history, albeit no less relevant to the philosophical roots of agricultural education. But there is a more specific history, and there are some specific lessons. These will be explored briefly in the following paragraphs.

The first school in America was an agricultural school. The students were a motley group of men, women, and children who had survived their first winter after landing at Plymouth. The teacher was an Indian, Squanto, who had agreed to teach them how to plant and to grow an indigenous crop--corn. He taught them how to fertilize each hill with fish caught in a nearby stream.

The elements of poignancy in this summary should not be overlooked. The classroom was an open field. The teacher was illiterate. The method was learning by doing. Because there was neither time for nor interest in frills, the teacher taught only the basics--survival or, more specifically, food production. All the students were foreign students and all, or most of them, did well in later life.

Much of the early history of agriculture is the history of fear of food shortages or famine. In many countries of the world, agricultural legislation began with the creation of a famine commission or a plan to eliminate food shortages. The U.S. Congress appropriated funds for the study of agriculture for the first time in 1838, when crop failures affected trade balances and required the importation of massive amounts of food. The beneficiary of much agricultural legislation and much agricultural development has been society and, except in the indirect sense, not the individual farmer.

But interest in agricultural education occurred long before the nineteenth century. In his second presidential message, George Washington proposed that Congress create a national agricultural university. Early in the nineteenth century, the existing universities began to teach agriculturally related subjects. Agricultural societies began to create small and rather isolated secondary schools, most of them influenced by the socio-economic stream of influence from England and thereby intent on improving but not changing the student's station in life.

The land-grant legislation initiated in 1857 was revolutionary, for it began a movement with the promise of changing the character of all American society. As with most changes, the Morrill Act was a reflection of an on-going change; it was not its prime mover.

For the first and only time in U.S. history, a multi-level educational reform had rejected the status-oriented elitism of the socio-economic stream of influence inherited from England, embraced the concepts of inquiry and freedom which were a part of the philosophical stream, and accepted the focus on individualism as well as on work values, which were a part of the theistic stream. The purpose of the legislation was to ensure that the best education would be available to all classes, including the sons and daughters of farmers and mechanics.

In retrospect, the Morrill legislation did much more than was expected of it. It initiated the creation of professional schools--engineering, education, medicine, pharmacy, law, and many others. Professional schools simultaneously became the centers of activity for the professions and stimulated their growth. The professions had fallen into disrepute after the revolution. Most members were linked with professional bodies in Europe, and most were suspected of being loyalists.

Judged by almost any standard, the land-grant universities were disappointing failures during the first 50 years of their existence. They were criticized by the well-established private universities for tolerating low standards. The most rigorous critics, however, were farmers and farm organizations. Farmers regarded them as unnecessary, and the National Grange actively opposed them for being impractical. The 50-year period from 1857 to 1907 was an uncomfortable period of time for land-grant colleges of agriculture.

By the end of the 50-year period, three developments, which had begun to create a climate of acceptance for colleges of agriculture, were well underway in rural America. First, there had been a rapid spread in the number of high schools, and many states had already begun the teaching of agriculture in rural schools. To accommodate this development, the Morrill Act was amended in 1907 to authorize the use of federally allocated funds for the training of teachers of agriculture. Second, colleges of agriculture had begun programs of agricultural research. Congress had responded in 1887 to requests to authorize federal allocations to agricultural research, and it became clear that agricultural colleges were seriously intent on addressing farmers' problems. Third, and most important to the acceptance of agricultural colleges, was the development, refinement, and implementation of the concept of service.

It was this concept which developed and continued as almost an esprit de corps among agricultural college faculty and their students. By the time Congress authorized an Agricultural Extension Service in 1914, the spirit of service was already imbedded in the fabric of the system; it had been a pivotal factor in winning the respect of farmers, farm organizations, and the general public. It continues to the present time. Agricultural faculty members at all levels are continuously available and responsive to their clients. This spirit of service was summed up in 1980 by Charles Funk, a Minnesota teacher of agriculture and the state's 1979 Teacher of the Year, when he said, "Farmers do not care how much you know until they know how much you care."

Why a Concern for Philosophy?

Every society tries to give its members, young and old, a framework for thinking, which will further the purposes of the society. Implicitly or explicitly, this framework becomes a part of the system of education. Education is one of a nation's most value-oriented activities.

Almost every educational decision is a value decision. There are few exceptions. The subjects to be included in the curriculum, for example, are merely means and never ends. The presence or absence of subjects in the curriculum is the consequence of value-oriented decisions--or it may be a mere accommodation to conventional wisdom, another kind of value choice.

Educational systems are highly preoccupied with organizing various categories of knowledge. There is a tendency to include some information at the expense of other information, and to emphasize some forms of instruction at the expense of others. Schools tend to deal with highly conventionalized forms of knowledge and with highly conventionalized approaches to scheduling, teaching, and counting the clients (enrollments). They therefore can be expected to reflect patterns of inclusion and exclusion, emphasis and subordination, which support their conventions. To the extent that such patterns tend to diminish the stature or the scope of subjects such as agriculture or its essential corollary--the spirit of service, it can be expected that schools exercise significant control over the way many individuals relate the importance of agriculture to the purposes of society. The previously mentioned revolution, which is now described as a land-grant college philosophy, was a change in values, a change in the conventional wisdom about both subject matter and the important role of service.

Philosophy does not contribute to knowledge by providing answers to complex questions. It contributes by offering assistance in raising the right questions. It helps in ascertaining how the canons of a field stand up under questioning, and it helps in framing the questions.

A concern for philosophy is a concern for standards, along with an interest in the type of inquiry which may be allowed to alter the nature of the standards. An interest in standards is also sufficient to differentiate between ends and means and to be threatening to the various means-related activities which are pursued as if they were ends.

How Is Philosophy Recognized?

A beliefs-and-values view of philosophy in individuals is recognized by expressions or manifestations of cultural norms, including religious behavior, and either adherence to or departure from such norms. It is further recognized by evidence that choices made are a reflection of beliefs or values held. The beliefs-and-values view of philosophy is difficult to distinguish from ideology, where beliefs are unchanged by any evidence to the contrary.

A standards view of philosophy is recognized by expressions and manifestations of an unending search for appropriate relationships between ends and means and by efforts to test whether the standards employed can be adequately verified. Those with a standards view will recognize that their standards are rarely neutral. Standards which exert pressures on dogma or on unquestioned traditions will sooner or later have adversaries as well as allies.

The Importance of Philosophy to Teacher Education in Agriculture

At best, a philosophy for teacher education in agriculture is a framework for thinking about, and acting on, the goals and the ends-means relationships in agricultural education. It is a framework derived from, and continuously influenced by, standards and values which are formed by inquiry, inquiry which leads to empirical verification and consistency with reality. Philosophy, therefore, is at the heart of the enterprise. It requires a consideration of the destiny of individuals, groups, and society itself. It is concerned, therefore, with the concept of destinations (ends) and the appropriate choice of routes (means) which may be available, or which may be made available.

A way of viewing such a framework is to create categories of goals and purposes. In doing so, it is necessary to label the categories as ultimate, central, essential, and instrumental, for example. What is most important to emphasize here is that the choice of categories is itself a value decision, a decision which has led the discipline of philosophy to create its various schools of thought--essentialism, positivism, pragmatism, etc. It may be interesting to see how a philosophy for agricultural education may fit into these categories. Since categories are necessary for a framework, one will be used in the following paragraphs with only the intent to show the strength of structure, not to show labels.

In the broadest sense, agricultural education seeks to promote both the highest possible levels of individual fulfillment and the best possible use of human resources. On one hand, agricultural education is an on-going effort to improve the ability of individuals to survive, adjust, and advance through multiple relations with a dynamic agriculturally related environment. On the other hand, it is an on-going effort to improve the capacity of this environment to use an expanding array of human capabilities for achieving desirable ends.

These seemingly disparate goals need to be brought into balance by constantly testing the assumption that individual goals and environmental or community goals are essential dimensions of the common goal of human resource development.

With such a dual-goal structure it would be possible, and perhaps desirable, to separate individual goals from the goals of environmental or community improvement. Such might seem reasonable if an individual's goal were to escape from the environment, or if the goal of improving the environment could not be realized. Historically, such situations have produced refugees, mass migrations, and revolutions. The problem has been the lack of proximity of resources to people--physical, intellectual, esthetic, spiritual resources-- and the conditions needed for self-determination.

With an unswerving commitment to clients, agriculture teachers might accept the view, therefore, that the important goals are connecting; they involve the achievement of individuals and the capacity of the environment to absorb such achievement. It would be unfortunate if teachers overlooked some of the important ways to improve the environment by inattention to justice and freedom, for example. Justice is not just the business of the courts. More important, it is the business of democracy's workshop, the schools. Justice involves the system for distributing awards, rewards, status, opportunities, and penalties. All teachers give and receive lessons about justice. Although not the same thing, freedom too may be discussed, restricted, defended, and applauded without being exercised. It is the exercise of freedom which increases its value and improves the capacity of the environment to utilize human resources.

Perhaps the most essential framework for viewing the importance of philosophy is the concept of professionalism as seen in the mutual sharing of standards and values. Ordinarily this concept contains five dimensions. These are:

1. A commitment to the needs of clients, including the full range of clients and client systems served by agricultural education. Families, communities, organizations, and institutions are not excluded.

2. A commitment to substance of subject matter requiring professionals to seek improvement in the base of knowledge available to clients.

3. A commitment to practice of the procedures for connecting the needs of clients to the base of substantive knowledge for realizing the ends sought by the connection.

4. A commitment to standards for increasing the knowledge of clients, for improving the quality and relevance of the knowledge base, and for elevating the quality of practice.

5. A code of ethics for publicly declaring the extent to which members of the profession will offer each other mutual support in accepting professional obligations.

Summary

Philosophy is important to teacher education in agriculture when those in the profession understand that it is not the substance of armchair speculation nor is it an activity to be engaged in when there is little else to do. It is central to the decisions which must be made about why agriculture teachers engage in specific activities, promote particular organizations, emphasize certain teaching methods, and provide various kinds of services to individuals and communities. It is important when there is an orderly standards-and-value basis for making decisions and for reraising "why" questions.

CHAPTER 17

Current Issues and Future Outlook

Edgar A. Persons
UNIVERSITY OF MINNESOTA

C. Cayce Scarborough
AUBURN UNIVERSITY

Identifying the current issues, indicating trends that appear to be shaping our programs, and suggesting a possible future for teacher education programs in agricultural education are discussed in this chapter. Also considered is how to have these matters on records for those who may wish to examine them at some future date.

The study of an educational situation would be incomplete without consideration of the circumstances surrounding the particular educational program. For example, the study of any vocational program during the 1960's and 1970's would be incomplete without consideration of the Vocational Education Act of 1963 (17.1) and its subsequent amendments. Likewise, a study of the 1963 Vocational Education Act would be incomplete without a study of the socio-educational situation prior to 1963 and its impact upon that legislation. The President's Panel of Consultants on Vocational Education (17.2) identified many of the issues influencing the legislation.

Identifying and clarifying the current issues which have an impact upon an education program is essential to understanding the options available and the reasons for selecting one option over another. Thus, this chapter should serve as such a means. Subsequent years may indicate whether the options selected were those most appropriate.

Issues in Teacher Education: Historical Perspective

Since the Smith-Hughes Act in 1917, teachers of agriculture have been required to earn a college degree in agriculture from an approved (registered) program (17.3). The approved program has been part of the certification process in the states. Thus, one of the first steps in teacher education in agriculture was the development of an undergraduate program at agriculture colleges for prospective teachers of vocational agriculture.

One of the first issues was whether this teacher education program should be based in the college of agriculture or in the college of education. This issue was settled by each state. At one time, the location of these teacher education programs in the United States was about equally divided between colleges of agriculture and colleges of education. Today the issue continues to be debated, with arguments citing advantages of being located in the college of agriculture or in the college of education.

At any rate, teacher educators in agriculture have worked with both colleges in developing programs to prepare teachers of vocational agriculture adequately. The goal has been to prepare teachers who are competent in agricultural subject matter and also effective as teachers of agriculture for inschool youth and adults.

Another issue that has been important is who controls the content of the teacher education programs for prospective teachers of vocational agriculture. Obviously, the college curriculum is decided by the college. However, the state department of education in each state has a large say in the certification of all public school teachers. Basically, certification standards are the efforts by states to assure competent teachers in the public schools. Thus, an "approved program" means that the program meets the standard set by a state department of education. At times, these standards have included specific listing of courses. Such specificity can conflict or at least complicate requirements of the college.

A related issue that has possibly become more important in recent years is the extent of specialization in agriculture. For example, the professor of agronomy, who has a Ph.D. and who has spent his professional life in one specialized area of agriculture, has difficulty in understanding how an individual with a B.S. degree can be competent in "all" areas of agriculture. This question is still being debated in these days of more specialization of teaching areas in agriculture. Some states have changed their certification requirements to reflect this specialization, while others still retain general certification requirements.

An issue facing leaders in agricultural education from the beginning is the extent to which the job of the teacher of agriculture influences the undergraduate program for the prospective teacher. This question has been prominent in all areas of teacher education in recent years through competency/performance-based teacher education programs. Sometimes this job emphasis is called "field orientation." The extreme view here is that the teacher education program at the teacher-training institution should be developed around the job of the teacher of agriculture in the field. Put another way, all efforts in the teacher education program should be directed toward developing a competent beginning teacher of agriculture. The major question about this approach is that it could lead to supporting the status quo, ignoring the need for innovation and new ways of developing a more effective local program. The alternative approach to competency/performance-based teacher education programs might be to emphasize developing a more effective local program rather than maintaining the local program as it is now being conducted in the community. Agricultural educators were discussing the CBTE/PBTE approach to teacher education as early as 1954.

An associated issue, which developed in a number of states from the two preceding issues, is the involvement of the state supervisor in the teacher education program. In some states, the state supervisor of agricultural education has assumed a prominent role in several areas of the teacher education program. Apparently, this "authority" was based upon his/her having field orientation (knowledge of what was needed in the field) and his/her being a representative of the state. This position was given some credence through the financial support of teacher education, from the budget for vocational education which was controlled by the state director of vocational education (frequently 50 percent of the total budget). The long-standing arrangement of putting vocational funds into teacher education programs at the colleges on a reimbursement basis was used as further support for involvement of state supervisors in the program. Such joint involvement had the advantage of developing cooperative working relationships between supervisors and teacher educators for the improvement of both programs. However, in some cases, there were definite indications that the state supervisor was to be "in charge" of all agricultural education programs in the state, including teacher education. In 1967, the USOE Program Officer for Agricultural Education, N. H. Hunsicker, supported this position in a speech entitled "New Challenges in Teacher Education and Supervision."

> Joint action by supervisors and teacher educators seldom occurs unless there is one recognized leader of the two groups. This leadership has long been recognized as the responsibility of the Head State Supervisor of Agricultural Education. To perform this role the state supervisor must exhibit tolerance, diplomacy, understanding of the program, and considerable persuasive ability. He must honestly seek and appreciate suggestions. It behooves the supervisor of the teacher education staff not only to accept the leadership of the Head State Supervisor, but to insist upon it (17.4).

Finally, what constitutes appropriate inservice education for teachers of agriculture has long been a problem. Teacher educators get involved in at least two ways. First, they should have something to offer toward the further professional development of the teacher. Second, if the teacher desires college credit and/or a graduate degree, teacher educators may represent the college.

Determining the extent that graduate courses toward a degree can also serve as inservice education needed by the teacher to do a more effective job of teaching in the local situation is difficult, as is determining the extent to which teacher educators should be involved in short workshops and other non-credit programs for professional improvement. In times of accountability, these questions must be resolved.

This brief historical review of issues in teacher education in agriculture serves to illustrate the evolutionary nature of issues. Some issues spring from the adoption of state and federal management policies, while others emerge because of the changing nature of the various agricultural professions. Although the long-standing issues have not been resolved, those specific areas about which there is current concern should be examined. They are not unrelated to the historical issues, but rather they have surfaced as new facets of the same general problems.

Current Issues in Instruction

Because each state and each teacher-training institution have numerous issues that are of importance to their teacher education program, only a core of issues that are universal to the profession are discussed here. These issues have been divided into four categories: general, secondary, post-secondary, and out-of-school youth and adult, recognizing that states emphasize different aspects of each of these phases of agricultural education.

General Issues

General issues cut across teacher education for all levels of vocational agricultural education. To suggest that some of these broad issues are more important than others would be a failure to recognize the vastness of the country and the diversity of teacher education programs. Many of these issues are not necessarily new; some of the problems and concerns of vocational teacher education in agriculture have been around for a long time.

<u>Teacher Supply</u>. Across the country the supply of agriculture teachers has failed to keep up with the demand. Although Chapter 5 addressed the subject of teacher recruitment, there are other facets of teacher supply that should be examined. Not only must efforts be made to recruit new teachers, but also attention must be directed to the problem of retaining those who enter the teaching field.

There have been numerous studies of reasons why teachers leave the profession. The reasons given are many, but they fall into the general categories of dissatisfaction with aspects of the job, inadequate financial rewards, lack of administrative support, disillusionment about youth, and few advancement opportunities. One of the factors that has not been widely recognized is the hypothesis that many who begin teaching do so with the idea that teaching is but a short-term steppingstone to a planned second, longer-term career in some other agricultural endeavor. If this hypothesis is indeed fact, and there is strong evidence that it is, then the profession needs to recognize the place of teaching in the career ladder of beginning teachers and take steps to use this factor in the recruitment process.

When potential teachers plan short-term teaching careers, there are implications for the professional organizations which represent teachers and teacher education. It is more difficult for individuals to build a strong allegiance to the profession when their long-term goals are in non-teaching fields. Thus, professional organizations must work harder to build programs that will meet the needs of short-tenured professionals, and they must concentrate some effort on programs that will extend short-term teaching careers into long-term ones through a planned system of incentives and rewards.

<u>Teacher Preparation Standards/Certification</u>. There has always been a question of what constitutes the proper mixture of professional education, technical subject matter, and basic learning in the arts and sciences. At the same time, teacher preparation has been subject to pressures to include additional training for working with special populations, improving human relations, and meeting specific state requirements about state history and government. The time a potential teacher has to prepare for a teaching role is

generally finite. The addition of mandated instruction in non-traditional study areas requires an examination of the instruction in education, technical agriculture, and arts and sciences, instruction that must be foregone to allow for the inclusion of the prescribed studies.

As states and institutions consider the addition of mandated courses of instruction in teacher preparation, it is essential that they maintain some semblance of standards for teacher preparation. At issue is enough uniformity in teacher education programs so that teachers can maintain some degree of job mobility between states and can meet some national standards for adequate preparation.

At issue also is the question of the specialization of initial teacher preparation. Should teachers be prepared for one or more of the teaching taxonomies, or should preparation be more general to allow teachers to function at a beginning level in any of the taxonomies? Recognizing that there are differences between states in the diversity of local vocational agriculture programs, it is also important to recognize that restrictive certification complicates the problem of teacher placement and has an impact on the mobility of teachers from state to state. In any event, a national standard should be used to guide such certification so that program quality can be assured, regardless of the specificity of the training/certification program.

Work Experience for Teachers. If vocational agriculture is to remain vocational, it seems logical that some form of work experience is necessary. There will continue to be differences in the work experience requirements for teachers at the secondary, post-secondary, and adult levels. The question is not one of the importance of work experience, but rather it is one of what constitutes adequate experience and how we can measure it.

Should work experience be determined in terms of the time spent in an agricultural work situation, or would some measure of the competencies required be more appropriate? It is a question of quantity or quality. If a competency/performance-based curriculum is to be used to guide teacher education, then it may be necessary to work toward competency-based work experience. To do so would require a clear definition of the purpose of work experience and the objectives that work experience would be expected to satisfy. The current practice of requiring vocational teachers to have completed some specified amount of work experience prior to teaching is a practice subject to inquiry, since it has not been adequately tested by research.

Financing Teacher Education. Operating programs in an arena where public funds for teacher education are increasingly restrictive will require further examination of the role of the federal government and state vocational education agencies in sharing the burden of teacher education with educational institutions. Some states, using a combination of flow-through federal funds and appropriate state funds, already share in the task. Other states have abrogated a sharing responsibility or have designated the state-federal share for restrictive uses. Which approach is best? Should institutions be solely responsible for preservice teacher education, with state and federal resources limited to the support of inservice education? Or, should the costs of all programs be a shared responsibility? To answer these questions requires an examination of the purpose of federal support for vocational education and the impact that support for teacher education may have, compared to a like expenditure for local program maintenance. It is clear that funds used only to supplant institutional dollars would have little impact, but funds spent to augment current expenditures could have major influence on the quality of outcomes from teacher education. Such augmentation may provide the needed incentive for increased recruitment and may allow curriculum modifications that address the specific needs of persons with long-term career goals outside teaching. Allowing for needed staff and resources to work with beginning teachers may be the catalyst needed to cause teachers to extend their tenure in teaching. Such extension may help teachers who have gained additional experience to do a more professional job, and it would reduce the number of replacements needed.

Secondary Issues

The general issues previously described all apply to the preparation of teachers for secondary schools. There are some issues, however, that are of particular importance to the secondary teacher.

Nature of the Secondary Curriculum. In order for the preparation of secondary teachers to be functional, there must be some attention given to the curriculum they are expected to teach. It may have been adequate in the past to provide general guidelines for what should be taught, leaving the content details to the discretion of the teacher. Agricultural specialization and advancement of technology now provide such an awesome cafeteria of possible subject matter that beginning teachers could benefit from more definition of core curriculum content. The questions to be raised include: What proportion of the total curriculum should be designated as a standard core? Should there be differentiation in core content between urban and rural schools? What degree of specificity should the curriculum materials contain? These questions are becoming even more crucial in light of a move away from the year-long courses in vocational agriculture. As students have options for quarter, semester, and trimester length enrollments, it becomes more difficult to provide a curriculum that both addresses the basic tenets of agriculture and gives continuity and sequence in instruction. Teacher education, at both the preservice and inservice levels, must provide effective instruction on how to cope with the fragmented approach to agriculture. Since many teacher educators have not worked in the public schools under those conditions, and therefore have little experience from which to draw, careful study of curriculum organization for secondary schools is essential if prospective teachers are to be prepared for a diversity of local program organization schemes.

The Role of Supervised Occupational Experience Programs. Vocational agriculture has evolved beyond the farming program as the only source of occupational experience for students. The expansion of vocational agriculture to encompass the totality of agriculture/agribusiness has dictated the broadening of the supervised occupational experience concept. The broadened concept has brought with it new challenges for teacher preparation. The organization and management of experience programs for clientele with diverse occupational goals has made preparation for supervision more difficult. The relatively standard procedures for supervising on-farm experiences had to be enlarged to encompass the diversity of occupations and the many state and federal laws that apply to this diversity. In the process, the idea that all students in vocational agriculture had to have some form of occupational experience has been lost. Many of the students now enrolled have only inschool instruction. At issue is to what degree supervised occupational experience plays a role in the total education of the student and whether vocational education can be truly vocational when the experience element is ignored. While almost all teacher educators would subscribe to the idea that occupational experience is essential, operationalizing the idea in local programs is another matter.

The occupational experience idea is further complicated by the cafeteria style of program delivery that permits students to sample various aspects of agriculture without making a long-term commitment to prepare for an agriculture career. There is a need to examine the alternative ways in which students can engage in some form of supervised experience, given the short exposure some have to instruction in vocational agriculture.

The task for teacher education is to lead the way. Researching alternative occupational experience strategies is a place to start. The introduction of suitable, manageable alternatives to job placement for short-term vocational agriculture students may be a beginning. When used with well-designed studies which deal with the impact of such alternatives on the competency levels of students, this may result in both beginning and experienced teachers having a renewed interest in re-establishing some form of occupational experience as an integral part of vocational agriculture education.

Work Experience for Teachers. If supervised occupational experience programs are an integral part of vocational agriculture at the secondary level, it logically follows that these programs are also important for teachers. The experience background of those preparing to teach agriculture is more varied than was once the case. With more urban youth preparing to teach, the problem of providing and requiring suitable work experience is more prominent. Students with agricultural backgrounds are more likely to come from farms or businesses where their experience has been in specialized enterprises or business endeavors. At issue is whose job it is to provide the necessary experience. Is obtaining experience the only responsibility of the student, or is it the responsibility of the teacher-training institution to arrange for, and supervise, an appropriate experience? Also unanswered is the question of what constitutes adequate experience. If it is the task of teacher-training institutions to intervene in the experience-gaining process, there will have to be provisions for staffs to organize, supervise, and evaluate the experience. It is an issue related to the financial resources of the institution and to the philosophical question of responsibility for the task.

The Teacher's Role. Examining the studies of why teachers leave the profession would prompt one to ask, "Is the job too big?" One of the tasks of teacher education would appear to be assisting beginning teachers in building a fence around their job. To do so requires the establishment of priorities and a logical rationale for determining what teacher tasks are more important than others. There is some indication that teacher education has not done an outstanding job of preparing teachers to make priority decisions. Just as the core curriculum idea can assist teachers in determining what should be included in the curriculum, a core activity itinerary can assist teachers in zeroing in on the important tasks for beginning teachers.

The issue is What constitutes the core of reasonable expectations for the beginning teacher of agriculture? While there are no doubt differences between states and regions, some effort should be made to address the issue. It is inappropriate to model this expectation after the successful teacher with long tenure. Given the short tenure of a growing number of teachers, it is the successful beginner who must serve as the role model in defining the activity core.

Post-secondary Issues

Because post-secondary education in agriculture occurs at non-collegiate institutions, such as vocational-technical institutions, and at junior and community colleges, the training of teachers for such institutions is more difficult to describe. There are, however, some concerns that are universal, regardless of the type of post-secondary institution. Two of these concerns are addressed.

Experience vs. Pedagogy. It is generally true that the background required in educational disciplines for initial employment is less rigorous for the post-secondary teacher than for the secondary teacher. The major emphasis is often placed on preparation in a subject matter area, with some institutions stresing relevant work experience more than formal education. The question is one of the appropriate mixture of knowledge and training in education and experience or study in the technical disciplines.

For teacher education, the problems are twofold: (1) What should be included in the pedagogical training for such instructors? and (2) What is the most effective procedure for delivery? The problem of what should be taught is further complicated by the diverse backgrounds of the prospective enrollees, while the problems of delivery are hampered by the limited number of potential clientele.

To examine the first question requires some assessment of the minimum competencies that teachers should possess to function in a classroom setting with young adults. The following are general areas in which it can be assumed that all teachers, regardless of level, should possess some knowledge of skill.

1. Psychology of how people learn.

2. Skill in organizing subject content in a logical, systematic plan.

3. Methods of teaching that are appropriate to the subject content and that incorporate the accepted principles of learning.

4. Procedures in evaluating the progress of students and in assessing the overall performance of the institutional program.

In addition, some post-secondary teachers may find that their assignment of supervising work experiences of students, operating laboratories for specialized study, or performing as advisors and counselors of students requires additional educational competencies. Should instruction to develop the needed competencies occur prior to employment, or should it occur as an inservice activity? That is the issue. To assume that instruction or experience in an agricultural subject matter area is adequate preparation for initial employment as a post-secondary teacher implies that the skills in teaching are of secondary importance. To insist that post-secondary instructors have some pedagogical training prior to employment strikes a more favorable balance between knowledge of teaching and knowledge of subject matter as a prerequisite to adequate job performance.

Delivery of pedagogical instruction to post-secondary teachers is another matter of concern. Since there are generally fewer teachers or potential teachers at the post-secondary level, and the turnover rate is assumed to be lower than for secondary teachers, the number of teachers to be served each year by each teacher-training institution is likely to be small. Thus, teacher education institutions need to be innovative in their approach to program delivery.

If pedagogical training is done on an inservice basis, special care must be taken to ensure that the location of institutions and the time at which it is offered have maximum utility for post-secondary instructors. Assessment of needs and careful plannng as to time and location are essential. Courses or training sessions offered in a location central to the potential enrollees in short, intensive time periods are most likely to appeal to post-secondary instructors.

The use of individualized instruction for post-secondary teacher training is also an alternative. When the number of post-secondary instructors does not warrant an organized group approach, a teacher trainer can make effective use of individualized instructional material and concentrate his/her time on the supervision of, and counseling with, the teacher. Such an approach, however, ignores the benefits of group interaction, which include the insight teachers can gain about teaching from each other.

Keeping Instruction and Curriculum Current. While the practice of hiring post-secondary teachers with adequate relevant work experience assures some practical knowledge of the problems of the field, one must recognize that such knowledge can soon become outdated. In agriculture, where technological change is rapid, teacher obsolescence can be a serious problem. The question is: How does teacher education structure education programs and experiences so that the teacher can keep pace with technological change?

It should be obvious that considerable attention must be directed to the subject matter field. There are various approaches that may be taken. By organizing workshops, industry-training sessions, and inservice courses, teacher education institutions can provide opportunities for teachers to update and improve their skills. They should also give careful consideration to making opportunities available whereby individuals can acquire skills to return to the work force. Employing institutions can grant work leaves, sabbatical leaves, and industry exchange programs for post-secondary instructors. There are multiple benefits. Not only does the instructor get reacquainted with the current technology of the industry, but such activity often provides students with a greater awareness of placement opportunities, and it triggers in them an assessment of the adequacy of their preparation to become successfully employed.

Out-of-School Youth and Adult Issues

There are probably more serious issues to be considered in the organization, financing, and supervision of instructional programs for out-of-school youth and adults than there are in the training of teachers. Yet, there are two central issues that stand out as concerns for teacher education: (1) providing suitable, initial preparation for out-of-school agriculture instructors and (2) maintaining instructor competence through inservice training. Discussion of these issues is hampered by the lack of a uniform definition of out-of-school instruction and by the variety of practices that states follow in implementing such activity.

In many states, the operation of the out-of-school program is considered an add-on activity for the secondary instructor. In others, this responsibility is relegated to full-time teaching personnel. Still others have a combination of some full-time instructors and some part-time instructors who teach adults in addition to secondary students.

The issues addressed in this chapter are related to the training of instructors who have a major or full-time responsibility for instruction in agriculture for out-of-school youth and adults.

Providing Preservice Preparation. To provide adequate preservice preparation necessitates recognizing that the job of the out-of-school instructor requires a broader range of competencies than that of the secondary instructor or the post-secondary instructor who is teaching in a narrow discipline field. Teachers of out-of-school youth must be able to instruct young farmers in how to get established in farming and in how to develop rural leadership. Teachers of adults must be able to instruct adults in how to manage established businesses. Both groups of teachers must have a thorough knowledge of agricultural technology and an understanding of the unique aspects of adult learning.

Therefore, the question arises as to whether or not teaching out-of-school youth is an entry-level profession or whether or not such teachers should first have experience as teachers of agriculture at either the secondary or the post-secondary level. Irrespective of how the question is answered, persons preparing to teach agriculture to out-of-school youth and adults need special competencies. Teacher educators have the task of organizing the program to provide this preparation to teach. Since many of the special competencies associated with establishment in farming and whole farm management are outside the realm of agricultural education pedagogy, such organization will require a close liaison with other departments at the teacher-training institutions.

Another question is whether or not every teacher education institution should provide special instruction in teaching out-of-school persons. It may be more practical to organize a consortium of institutions, in which one institution in a region develops the special expertise for teacher preparation. Such a consortium is now operated by the Committee on Institutional Cooperation (CIC),* which includes the Big Ten universities and the University of Chicago, for specialized educational opportunities. Agricultural teacher preparation for out-of-school instruction could become a thrust of this type of consortium.

Providing Inservice Education for Out-of-School Instructors. Because adult and out-of-school education usually involves a great deal of on-farm instruction, the problem of retaining technical competence is generally less severe than with other levels of instruction. By the same token, the specialized skills in management and the pedagogy associated with effective instruction need constant updating. Here the issues center around the organization and management of the inservice activity. If the preservice preparation is relegated to a consortium, the expertise for offering inservice may be limited. Teacher trainers may find their role defined as facilitators of

*The Committee on Institutional Cooperation is located at 820 Davis Street, Suite 130, Evanston, Illinois 60201.

inservice education. Organizing the inservice activity in courses or workshops using qualified personnel outside agricultural education may be an appropriate strategy when teacher-training staffs lack expertise in the technical disciplines. As with post-secondary instructors, determining the time and location to maximize utility is a key consideration. Teacher education must assume a leadership role in providing the needed inservice opportunities in this area.

Future Outlook

What's ahead for teacher education in agricultural education? Any responsible prediction of the future of any program is based upon certain assumptions--the basic assumption being that there will be a program.

The fact that there is no identification of teacher education as a requirement in present national vocational legislation with provision for financing same raises some question. This is in contrast to the Smith-Hughes Act of 1917 (17.5), where teacher education was mandated. Since those days, teacher education has slipped from being a mandate to being an ancillary service, and it is now missing entirely in the language of the legislation. Apparently, insofar as national legislation is concerned, teacher education is optional. There appears to be an assumption by some that the desired goal of vocational education's being readily available in all communities to all people can become a reality without teacher education.

Thus, it seems that the immediate future of teacher education in agriculture will depend upon the ability of teacher educators to "sell" their program as a responsibility of the university. The long-range future of the program may well depend upon the future job parameters of the teacher of vocational agriculture. In some states, this has already changed from a year-round continuing program to a 10-month teaching job. A major question having indirect influence is whether students will continue to enroll in agricultural education to prepare to teach in a 10-month program. One possible option already operating in several states is to view agricultural education as a program for developing leaders in <u>any</u> and <u>all</u> agricultural professions, not just for teaching vocational agriculture. This has, in fact, already happened to some extent, as indicated by the wide range of leaders in agricultural professions who started out in agricultural education.

What, then, does the future hold? The current issues in teacher education are not new. It is likely that the issues of the future will be largely manifestations of changes in education that are already in progress. Change in education is evolutionary rather than revolutionary. Given that change is imminent, what issues are likely to surface?

It is unlikely that the basic tenets of vocational education in agriculture will undergo serious alteration. The basic concepts have stood the test of time without major modification. The last major change occurred in 1963 with the expansion of vocational agriculture to encompass off-farm as well as on-farm agricultural occupations. What will likely change is the tools with which educators work, the environment in which education occurs, and the priorities by which education is directed, at the various levels of instruction. As with any future projection, these changes must be viewed according to the premise that "all other things are equal."

The Changing Environment

Demographic projections for the United States clearly indicate that there will be changes in the school environment in which vocational agriculture is offered. There are continuous and perceptible shifts in the regional populations of the United States. These shifts may be even further accelerated as the questions of energy availability and the quality of life have an impact on decisions about where to live.

Agriculture has not been immune to demographic change. Almost all sections of the country have experienced a decline in the number of farms. Less obvious has been a reduction in family size. Combined with reduction in farms, this has resulted in major losses in rural populations. The impact of these changes will be most severely felt during the 1980's as the declining number of children in schools reduces school size. With many rural communities already at minimum school size, further reductions will have dynamic effects: schools will be closed, subject matter offerings will be reduced, and the economic plight of educational institutions will become even more severe.

Declining school populations raise some important issues for teacher education. The effect that the decline will have on the number of teachers employed will depend upon the growth and expansion at the out-of-school youth/ adult and post-secondary levels. In order to maintain the present cadre of teachers, it will be necessary to shift some of the emphasis on the preparaton of secondary teachers to other levels. Considering that teacher-training institutions have been reluctant to deviate from secondary teacher preparation, the net result of such failure will be an overall reduction in teacher numbers and a decline in enrollments in agricultural education at the college level. Serious declines will result in reduced staffing, curtailed programs, and less capacity to respond to both the preservice and inservice needs of clientele.

The declining number of secondary graduates will accentuate an already severe recruitment problem. This problem will be amplified by expanded alternatives for secondary graduates, as other post-secondary programs build a capacity to attract agriculturally oriented students. If labor market predictions for the 1980's are true, there will also be increased competition from industry where workers, both trained and untrained, will be in short supply.

How will teacher education institutions cope with these changes? Will there need to be redirection in training programs for teachers? Only the hindsight of the 1990's will provide a clear picture of what happened, but institutions do have options for charting their own course. Diverse tactics such as forming consortiums, expanding the focus of teacher training to encompass other than secondary teacher preparation, and adjusting training programs to prepare teachers for roles in expanded but fewer vocational agriculture departments are but some of the options that must be exercised if teacher training is to remain a viable enterprise.

Educational Tools

Teachers of the future will have at their fingertips a wide variety of sophisticated teaching tools. The novelties of the 1960's, which became exemplary in the 1970's, will be commonplace in the 1980's. Probably foremost among the new tools will be additional applications of computer technology to the teaching process. It is conceivable that textbooks as we now know them will be replaced by instructional material constantly updated and available through computer technology. When it is combined with the capability to contrive simulations, utilizing all the factors of production and control that are significant to both on-farm and off-farm agricultural enterprises, the possibilities for instruction are unlimited. Developing courses of study and carricula for agricultural instruction will be guided by a vast storehouse of information that can be retrieved, edited, and organized to suit the needs of any agricultural community. The technology for such activity already exists. Application of the technology is under development, and adaption of the technology will soon follow.

What are the implications for teacher education? It is likely that computerized teaching tools will be developed in the private sector. Teacher education is, and will continue to be, an obligation of the public sector. If teacher education institutions are to train prospective and inservice teachers at the secondary, post-secondary, and out-of-school youth/adult levels who know how to get maximum utility from the emerging technology, they will have to form a partnership with the private sector. If teacher education institutions fail to seize the opportunity to work with the private sector in this development, there will be two primary systems of education in agriculture--a private system with the latest instructional technology and a public system without benefit of the modern techniques for instruction. This prediction does not imply that one system is better than the other, for there are advantages to both systems. It does suggest that pooling resources of the public and private sectors in teacher education can be beneficial. The tools for education will change. The issue is: Should teacher education be a full partner in change or only a minor stockholder?

Nor are educational tools limited to computer technology. The video cassette player is already a common household appliance. Only the lack of a large catalog of information on the many aspects of agriculture limits its use in agricultural instruction. If triggered by the appropriate incentives, both the public and the private sectors will work at developing marketable instructional units on cassette videotape. Once these units are available, they will become part of the teaching strategies at all levels of agricultural instruction. At issue is: How will they be utilized in teacher education and how will teachers be prepared to maximize their utility?

Summary

Only two major future concerns have been addressed here: the changing environment for instruction and the tools of instruction. These are neither exhaustive descriptions of the issues of the future nor extensive samples of the range of possibilities. Instead, they are meant to call attention to the imminence of change. Changes and issues in teacher education in agriculture are evolutionary. While no one knows exactly what agricultural education of the future will be like, it is certain that it will be different.

Citations

Chapter 1

1.1. Logan, W. B. "Vocational Education: Facts and Misconceptions," Theory into Practice, Vol. 3, 1964, p. 161.

1.2. Charters, W. W., Jr. "The Social Background of Teaching," in N. L. Gage, ed., Handbook of Research on Teaching, Chicago: Rand McNally & Company, 1963.

1.3. Saylor, G. "Secondary Education-Development," in C. W. Harris, ed., Encyclopedia of Educational Research, 3rd ed., New York: Macmillan Company, 1960, p. 1238.

1.4. Hawkins, L. S., Prosser, C. F., and Wright, J. C. Development of Vocational Education, Chicago: American Technical Society, 1951, pp. 123-131.

1.5. Eddy, E. D., Jr. Colleges for Our Land and Time, New York: Harper and Brothers, 1956, pp. 113-148.

1.6. Kerr, W. J. "The Spirit of Land-Grant Institutions," address delivered at the forty-fifth annual convention of the Association of Land-Grant Colleges and Universities, Chicago, November 1931, and published by The University of Arizona, Tucson, 1961.

1.7. Elder, C. R. "Peoples' Colleges," in A. Steffereed, ed., After a Hundred Years, 1962 Yearbook of Agriculture, USDA, Washington, D.C., U.S. Government Printing Office, 1962, p. 15.

1.8. Lord, R. The Care of the Earth, New York: Thomas Nelson and Sons, 1962, p. 237.

1.9. Lane, C. H. "Contribution of the United States Department of Agriculture to Agriculture of Less Than College Grade, 1904-1917," in History of Agricultural Education of Less Than College Grade in the United States, Vocational Education Bulletin No. 217, Agricultural Series No. 55, U.S. Office of Education, Washington, D.C., U.S. Government Printing Office, 1942, p. 570.

1.10. True, A. C. A History of Agricultural Education: 1785-1925, Miscellaneous Publication No. 36, USDA, Washington, D.C., U.S. Government Printing Office, 1929, p. 353.

1.11. *Ibid.*, p. 356.

1.12. *Ibid.*, pp. 272-273.

1.13. Swanson, H. B. "Teacher Training in Agriculture," in R. W. Stimson and F. W. Lathrop, comps., *History of Agricultural Education of Less Than College Grade in the United States*, Vocational Education Bulletin No. 217, Agricultural Series No. 55, U.S. Office of Education, Washington, D.C., U.S. Government Printing Office, 1942, p. 516.

1.14. *Ibid.*, pp. 517-518.

1.15. Lathrop, F. W. "Trends in Vocational Education in Agriculture," in R. W. Stimson and F. W. Lathrop, comps., *History of Agricultural Education of Less Than College Grade in the United States*, Vocational Education Bulletin No. 217, Agricultural Series No. 55, U.S. Office of Education, Washington, D.C., U.S. Government Printing Office, 1942, p. 621.

1.16. True. *Op. cit.*, p. 320.

1.17. Greenwood, E. "Attributes of a Profession," in S. Nosow and W. H. Form, eds., *Man, Work and Society*, New York: Basic Books, Inc., Publishers, 1962, pp. 206-217.

1.18. McGlothlin, W. J. *The Professional Schools*, New York: Center for Applied Research in Education, Inc., 1964, pp. 1-31.

1.19. Stinnett, T. M. *The Profession of Teaching*, New York: Center for Applied Research in Education, Inc., 1962, pp. 1-15.

1.20. Beggs, W. K. *The Education of Teachers*, New York: Center for Applied Research in Education, Inc., 1965, pp. 47-61.

1.21. Conant, J. B. *The Education of American Teachers*, New York: McGraw-Hill Book Company, 1963, pp. 209-218.

1.22. Stiles, L. T., *et al.* *Teacher Education in the United States*, New York: Ronald Press, 1960, pp. 290-291.

1.23. *Ibid.*, pp. 95-115.

1.24. *Directory of Agricultural Teacher Education Personnel in the United States, 1966-67*, American Association of Teacher Educators in Agriculture, October 1966.

1.25. Russell, J. D., *et al.* *Vocational Education*, Staff Study No. 8, prepared for the Advisory Committee on Education, Washington, D.C., U.S. Government Printing Office, 1938, pp. 171-199.

1.26. Martin, W. H. "Operational Interpretation of the Smith-Hughes Act As Reflected in the Writings of Teacher Educators in Vocational Agriculture," unpublished Ed.D. dissertation, University of Illinois, Urbana, 1953, pp. 111-119.

1.27. Standards for Quality Programs in Agricultural/Agribusiness Education, Agricultural Education Department, Iowa State University, Ames, 1977.

1.28. Stimson, R. W. "Home Project Teaching and Related Educational Developments," in R. F. Stimson and R. W. Lathrop, comps., History of Agricultural Education of Less Than College Grade in the United States, Vocational Education Bulletin No. 217, Agricultural Series No. 55, U.S. Office of Education, Washington, D.C., U.S. Government Printing Office, 1942, pp. 591-602.

1.29. Axelrod, J. "New Patterns of Internal Organization," in L. Wilson, ed., Emerging Patterns in American Higher Education, Washington, D.C.: American Council on Education, 1965, pp. 41-42.

1.30. Conant. Op. cit., pp. 56-72.

1.31. Stratemeyer, F. B. "Perspective on Action in Teacher Education," the sixth Charles W. Hunt Lecture of the American Association of Colleges for Teacher Education, Washington, D.C., American Association of Colleges for Teacher Education, 1965.

1.32. Professional Workers in State Agricultural Experiment Stations and Other Cooperating State Institutions, 1966-67, Agricultural Handbook No. 305, USDA, Washington, D.C., U.S. Government Printing Office, December 1966.

1.33. For a state-by state list of staff positions and organization structure of state departments of education, see R. F. Will, State Education--Structure and Organization, O.E. 23038, Miscellaneous Publication No. 46, Washington, D.C., U.S. Government Printing Office, 1964.

1.34. Swanson. Op. cit., pp. 518-523.

1.35. Bueke, V. L., et al. Implementation of the Educational Amendments of 1976: A Study of State and Local Compliance and Evaluation Practices, Final Report No. 80-113, prepared for the National Institute of Education, ABT Associates, Inc., Cambridge, Massachusetts, December 1980, p. 248.

1.36. Axelrod. Op. cit., p. 51.

1.37. Caplow, T., and McGee, R. J. The Academic Marketplace, New York: Basic Books, Inc., Publishers, 1959, pp. 79-137.

1.38. Lear, J. "Who Should Govern Medicine?," Saturday Review, June 5, 1965, pp. 39-42.

1.39. Stiles, et al. Op. cit., pp. 260-288.

1.40. Agricultural Education for the Seventies and Beyond, report prepared by the Committee on Agricultural Education, Commission on Education in Agriculture and Natural Resources, National Research Council, and published by the American Vocational Association, Washington, D.C., July 1971.

1.41. Guiding Principles for Pre-service Training of Teachers of Vocational Agriculture, Teacher Education Committee, Agricultural Education Division, American Vocational Association, Washington, D.C., 1962.

1.42. Recommended Standards for Teacher Education, report of the Evaluative Criteria Study Committee, American Association of Colleges for Teacher Education, Washington, D.C., November 1969.

1.43. Teacher Education in Agriculture Guidelines, report of the Guidelines Committee of the American Association of Teacher Educators in Agriculture, presented and approved at the American Vocational Association convention, Portland, Oregon, 1971.

1.44. Standards for Quality Programs . . ., op. cit.

1.45. Dougan, J. E. "A Philosophy of Vocational Agricultural Education," statement approved by the Agricultural Education Division, American Vocational Association, December 9, 1975.

1.46. Definitions of Terms in Vocational and Practical Arts Education, prepared by the Committee on Research and Publications, American Vocational Association, Washington, D.C., 1954.

1.47. Knebel, E. H., and Richardson, W. B., ed. Terminology of Importance to Professionals in Agricultural Education, American Vocational Association, Washington, D.C. (in press).

1.48. McClay, D. R. National Agricultural Occupations Competency Study, U.S. Office of Education, Washington, D.C., U.S. Government Printing Office, May 1978.

1.49. Spalding, W. B. "Evaluation of Proposals for Change in Teacher Education," in Changes in Teacher Education, An Appraisal, report of the National Commission on Teacher Education and Professional Standards at the Columbus conference, 1963, and published by the National Education Association, Washington, D.C., 1964, p. 48.

1.50. "The Innovating Organization," Transaction, special supplement, Vol. 2, 1965, pp. 30-40.

Chapter 2

2.1. Woodlin, R. J. "Teacher Power Is Necessary to Teacher Education in Agriculture," The Agricultural Education Magazine, Vol. 49, No. 4, October 1976, p. 75.

2.2. Thompson, O. E. The Possible Dream, 1972 AATEA lecture, Danville, Illinois: The Interstate Printers & Publishers, Inc., 1972, p. 10.

2.3. Cox, D. E., and McCormick, F. G. "Supervised Leadership Development for Prospective Agriculture Teachers--One Approach," The Agricultural Education Magazine, Vol. 50, No. 8, February 1978, p. 186.

2.4. Newcomb, L. H. "The Design of Teacher Education Field Experience to Meet the Needs of a New Clientele," The Agricultural Education Magazine, Vol. 49, No. 4, October 1976, p. 76.

2.5. Andrews, D. W. "Teacher Education in Agriculture," The Agricultural Education Magazine, Vol. 50, No. 5, November 1977, pp. 112-113.

2.6. Cooper, E. L. "So That Is What Teacher Educators Do," The Agricultural Education Magazine, Vol. 47, No. 2, August 1974, p. 35.

2.7. Krebs, A. H. "Agricultural Education: Some Problem Issues and Predictions," The Agricultural Education Magazine, Vol. 42, No. 1, July 1969, p. 6.

2.8. Warmbrod, J. R. "Teachers Determine Program Effectiveness," The Agricultural Education Magazine, Vol. 42, No. 7, January 1970, pp. 163-164.

2.9. Dillon, R. D. "The Leadman for Inservice Education," The Agricultural Education Magazine, Vol. 45, No. 4, October 1972, p. 75.

2.10. Wright, B. R., and Lucas, T. M. "Is Teacher Education Up-to-Date?," The Agricultural Education Magazine, Vol. 42, No. 7, January 1970, p. 167.

2.11. Atherton, J. C. "In Tune with Reality," The Agricultural Education Magazine, Vol. 42, No. 4, October 1969, p. 84.

2.12. Peterson, M. J. Agricultural Education: Some Issues and Some Reactions, 1968 AATEA lecture, Danville, Illinois: The Interstate Printers & Publishers, Inc., 1968, p. 9.

2.13. Kitts, H. W. "Professional Improvement," The Agricultural Education Magazine, Vol. 43, No. 11, May 1971, p. 263.

Chapter 3

3.1. O'Kelley, G. L., Jr. "Programs of Teacher Education in Agriculture," in V. R. Cardozier, ed., Teacher Education in Agriculture, 1st ed., Danville, Illinois: The Interstate Printers & Publishers, Inc., 1967, p. 30.

3.2. Stimson, R. W., and Lathrop, F. W., comps. History of Agricultural Education of Less Than College Grade in the United States, Bulletin No. 217, Agricultural Series No. 55, U.S. Office of Education, Washington, D.C., U.S. Government Printing Office, 1942.

3.3. Bender, R. E. "The Program of Agricultural Education with Implications for Colleges of Agriculture and Natural Resources," in D. L. Armstrong, ed., Impact of Enrollments and Student Body Composition on Academic Programs, Design, and Delivery, RICOP report, Michigan State University, East Lansing, 1977, p. 430.

3.4. O'Kelley, Jr. "Programs of Teacher Education . . .," op. cit., p. 33.

3.5. Guiding Principles for Institutions Training Teachers of Vocational Agriculture, report of the Teacher Education Committee at the annual conference of the American Vocational Association, Buffalo, New York, August 1958.

3.6. Guiding Principles for Pre-service Training of Teachers of Vocational Agriculture, report of the Teacher Education Committee, Agricultural Education Division, American Vocational Association, Washington, D.C., 1962, pp. 6-8.

3.7. O'Kelley, Jr., "Programs of Teacher Education . . .," op. cit, pp. 38-39.

3.8. Teacher Education in Agriculture Guidelines, American Association of Teacher Educators in Agriculture, 1976.

3.9. Bender. Op. cit., p. 439.

3.10. Standards for Quality Programs in Agricultural/Agribusiness Education, Agricultural Education Department, Iowa State University, Ames, 1977.

3.11. Oren, J. W. An Analysis of Standards for Teacher Education Programs in Agriculture/Agribusiness, report of the AATEA ad hoc committee on the Legitimization of Program Standards and Guidelines, published by Mississippi State University, Mississippi State, 1977.

3.12. Cottrell, D. D., ed. Teacher Education for a Free People. Washington, D.C.: American Association of Colleges for Teacher Education, 1956, p. 56.

3.13. Agricultural Education for the Seventies and Beyond, report prepared by the Committee on Agricultural Education, Commission on Education in Agriculture and Natural Resources, National Research Council, and published by the American Vocational Association, Washington, D.C., July 1971.

3.14. Ibid., p. 30.

3.15. Ibid., p. 31.

3.16. Hamilton, J. B., and Huang, M. W. Resource Person Guide to Using Performance Based Teacher Education Materials, Columbus, Ohio: National Center for Vocational Education, 1975, p. 7.

3.17. Bender. Op. cit., p. 432.

3.18. Cox, D. E. "Preparation of Teachers of Vocational Agriculture: An Attempt at a Program Model," unpublished monograph, Department of Higher Education, The University of Arizona, Tucson, 1979, p. 16.

3.19. Cangelosi, J. S. "Competency Based Teacher Education: A Cautionary Note," Contemporary Education, Vol. 46, No. 2, 1975, p. 124.

3.20. Cox. Op. cit., pp. 17-18.

3.21. Hamilton and Huang. Op. cit., p. 3.

3.22. Achieving the Potential for Performance-based Teacher Education: Recommendations, PBTE Series No. 16, American Association of Colleges for Teacher Education, Washington, D.C., 1974, pp. 32-33.

3.23. Hamilton and Huang. Op. cit., pp. 5-6.

3.24. Ibid., pp. 17-20.

3.25. O'Kelley, Jr. "Programs of Teacher Education . . .," op. cit., p. 47.

3.26. Agricultural Education for the Seventies and Beyond, op. cit., pp. 35-36.

3.27. Bender. Op. cit., p. 435.

3.28. Cottrell, ed. Op. cit., p. 208.

3.29. O'Kelley, G. L., Jr. Agricultural Education: The Preparation of Teachers, Vocational Education Bulletin No. 295, Agricultural Series No. 78, U.S. Office of Education, Washington, D.C., U.S. Government Printing Office, 1962, p. 48.

3.30. Agricultural Education for the Seventies and Beyond, op. cit., p. 34.

3.31. Zurbrick, P. R. "An Approach to Curriculum Development," National Association of Colleges and Teachers of Agriculture Journal, Vol. 20, No. 4, December 1976, pp. 22-24.

3.32. Bender. Op. cit., p. 434.

3.33. Agricultural Education for the Seventies and Beyond, op. cit., p. 7.

3.34. Bender. Op. cit., p. 442.

3.35. Bender, R. E., and Crawford, H. R. "Extension Education Should Be a Part of Agricultural Education," and "Extension Education Should Not Be a Part of Agricultural Education," The Journal of the American Association of Teacher Educators in Agriculture, Vol. 17, No. 2, July 1977, pp. 2-8, 14.

Chapter 4

4.1. Guralnik, D. B., ed. Webster's New World Dictionary of the American Language, New York: Simon and Schuster, Inc., 1980.

4.2. Hirst, B. A. "The Components of Competency Based Vocational Education," American Vocational Journal, Vol. 52, No. 8, November 1977, p. 32+.

4.3. Gentry, G. B. "Whither Agricultural Education--in Measuring and Evaluating Pupil Growth?," The Agricultural Education Magazine, Vol. 9, No. 9, March 1937, p. 131.

4.4. McClay, D. R. _National Agricultural Occupations Competency Study_, U.S. Office of Education, Washington, D.C., U.S. Government Printing Office, May 1978.

4.5. _Vocational Education Amendment of 1968_, Public Law 90-576, Title I, Washington, D.C., U.S. Government Printing Office, 1968, p. 1.

Chapter 5

5.1. Woodin, R. J. "Teachers Key Men in Recruitment Drive," _The Agricultural Education Magazine_, Vol. 38, No. 12, June 1966, p. 272.

5.2. Craig, D. G. _A National Study of the Supply and Demand for Teachers of Vocational Agriculture in 1977_, College of Education, University of Tennessee, Knoxville, March 1978, p. 1.

5.3. Peterson, R. L. _Review and Synthesis of Research in Vocational Teacher Education_, National Center for Vocational Education, The Ohio State University, Columbus, 1973, p. 10.

5.4. _Vocational Education Amendments of 1976_, Public Law 94-482, Washington, D.C., U.S. Government Printing Office, October 12, 1976.

5.5. McClay, D. R. _National Agricultural Occupations Competency Study_, U.S. Office of Education, Washington, D.C., U.S. Government Printing Office, May 1978.

5.6. Lee, J. S., _et al_. _Procedures for Recruiting, Selecting, and Preparing Persons with Non-teaching Professional Degrees to Be Teachers of Vocational Agriculture/Agribusiness_, Department of Agriculture and Extension Education, Mississippi State University, Mississippi State, 1978, p. 3.

5.7. Craig. _Op. cit._, p. 6.

5.8. _Ibid._, pp. 2-4.

5.9. _Standards for Quality Programs in Agricultural/Agribusiness Education_, Agricultural Education Department, Iowa State University, Ames, 1977, pp. x, 1-13.

5.10. Lee. _Op. cit._, p. 12.

5.11. Green, H. H. "Perceptions of the Extent of Use and Effectiveness of Selected Practices and Procedures in the Recruitment of Agricultural Education Students," in J. D. McCracken, comp., _Summaries of Research and Development Activities in Agricultural Education_, Department of Agricultural Education, The Ohio State University, Columbus, 1976, p. 64.

5.12. Lee. _Op. cit._, pp. 10-11.

5.13. Annis, W. H., and Paul, N. L. Recruitment and Selection Survey of Teacher Education Institutions, unpublished survey, Occupational Education Department, University of New Hampshire, Durham, 1979.

5.14. Craig. Op. cit., pp. 18-21.

5.15. Annis and Paul. Op. cit.

Chapter 6

6.1. Long Range Planning in Agricultural Education, North Carolina State University, Raleigh, November 1957.

6.2. Good, C. V. Dictionary of Education, 3rd ed., New York: McGraw-Hill Book Company, 1973, p. 258.

6.3. Hamlin, H. M. "The Curriculum: General Education," in V. R. Cardozier, ed., Teacher Education in Agriculture, 1st ed., Danville, Illinois: The Interstate Printers & Publishers, Inc., 1967.

6.4. Standards for Quality Programs in Agricultural/Agribusiness Education, Agricultural Education Department, Iowa State University, Ames, 1977.

6.5. Teacher Education in Agriculture Guidelines, American Association of Teacher Educators in Agriculture, 1976.

6.6. Guiding Principles for Pre-service Training of Teachers of Vocational Agriculture, report of the Teacher Education Committee, Agricultural Education Division, American Vocational Association, Washington, D.C., 1962.

6.7. Standards for the Accreditation of Teacher Education, National Council for the Accreditation of Teacher Education, Washington, D.C., 1979.

6.8. Elam, S., ed. Improving Teacher Education in the United States, Phi Delta Kappa, Inc., Bloomington, Indiana, 1967.

Chapter 7

7.1. Bender, R. E. "The Program of Agricultural Education with Implications for Colleges of Agriculture and Natural Resources," in D. L. Armstrong, ed., Impact of Enrollments and Student Body Composition on Academic Programs, Design, and Delivery, RICOP report, Michigan State University, East Lansing, 1977, p. 434.

7.2. Peterson, M. J., and Torrence, A. P. "The Curriculum: Agricultural Subject Matter," in V. R. Cardozier, ed., Teacher Education in Agriculture, 1st ed., Danville, Illinois: The Interstate Printers & Publishers, Inc., 1967, p. 137.

7.3. Standards for Quality Programs in Agricultural/Agribusiness Education, Agricultural Education Department, Iowa State University, Ames, 1977, pp. x-4.

7.4. Bender. Op. cit., p. 438.

7.5. Ibid., pp. 437-438.

7.6. Standards for Quality Programs . . ., op. cit., pp. 1-16.

7.7. Bender. Op. cit., p. 437.

7.8. Standards for Quality Programs . . ., op. cit., pp. x, 1-23.

7.9. McCracken, J. D., and Yoder, E. P. Determination of a Common Core of Basic Skills for Agribusiness and Natural Resources, Department of Agricultural Education, The Ohio State University, Columbus, 1975.

7.10. McClay, D. R. National Agricultural Occupations Competency Study, U.S. Office of Education, Washington, D.C., U.S. Government Printing Office, May 1978.

7.11. McCracken, J. D., and Warmbrod, J. R. "Identifying and Assessing Technical Competence of Prospective Teachers," The Journal of the American Association of Teacher Educators in Agriculture, Vol. 17, No. 3, November 1976, pp. 1-2.

7.12. Ibid., p. 2.

7.13. Ibid.

7.14. Peterson and Torrence. Op. cit., p. 138.

7.15. McCracken, J. D. Status of Pre-service Teacher Education in Agriculture, Central Region, paper presented at Central States Seminar in Agriculture/Agribusiness Education, Chicago, February 3, 1976.

7.16. Loreen, C. O. "A Study of the Agricultural Education Curricula of Forty-nine Teacher Training Centers in the United States and Puerto Rico," The Agricultural Education Magazine, Vol. 6, February 1953.

7.17. Peterson and Torrence. Op. cit.

7.18. McCracken. Op. cit.

7.19. Standards for Quality Programs . . ., op. cit., pp. 1-4.

7.20. Oscar. Op. cit.

7.21. Peterson and Torrence. Op. cit.

7.22. McCracken. Op. cit.

7.23. Yoder, E. P., and Bender, R. E. Development and Implementation of Internship Programs in Agricultural Occupations for Present and Prospective Vocational Agriculture Teachers, Department of Agricultural Education, The Ohio State University, Columbus, 1976, p. iii.

7.24. Ibid., pp. 1-2.

7.25. Ibid., p. 2.

7.26. Bender. Op. cit., p. 435.

7.27. Yoder and Bender. Op. cit., pp. 5-6.

7.28. Evans, D. Guidelines for Agriculture 430 Internship, College of Agriculture, The Pennsylvania State University, University Park, 1975.

7.29 McCracken and Warmbrod. Op. cit., pp. 2-3.

7.30. Ibid., p. 3.

7.31. Ibid., p. 6.

7.32. Ibid.

7.33. Ibid., pp. 6-7.

Chapter 8

8.1. Stabler, E. "The Current Scene in Teacher Education," in E. Stabler, ed., The Education of the Secondary School Teacher, Middleton, Connecticut: Wesleyan University Press, 1962, p. 4.

8.2. Crosby, D. J. "Training Courses for Teachers of Agriculture," Yearbook of the United States Department of Agriculture: 1907, USDA, Washington, D.C., U.S. Government Printing Office, 1908, p. 212.

8.3. Teacher Training in Agriculture, Bulletin No. 94, Agricultural Series No. 20, Federal Board for Vocational Education, Washington, D.C., U.S. Government Printing Office, 1924, p. 54.

8.4. The Work of the Agricultural Colleges in Training Teachers of Agriculture for Secondary Schools, Circular No. 118, USDA, Washington, D.C., U.S. Government Printing Office, 1913, p. 4.

8.5. Barringer, B. E. Student Teaching in Agriculture, Bulletin No. 100, Agricultural Series No. 23, Federal Board for Vocational Education, Washington, D.C., U.S. Government Printing Office, 1925, p. 9.

8.6. Lattig, H. E. Practical Methods in Teaching Vocational Agriculture, New York: McGraw-Hill Book Company, 1931, pp. 1-2.

8.7. Teacher Training in Agriculture, op. cit., p. 2.

8.8. Taylor, R. E. "The Professional Teacher and the State Program," The Agricultural Education Magazine, Vol. 37, April 1965, p. 237.

8.9. Woodin, R. J. "Common Competencies for All Vocational Teachers," The Agricultural Education Magazine, Vol. 37, February 1965, p. 188.

8.10. Broudy, H. S. "Criteria for the Professional Preparation of Teachers," The Journal of Teacher Education, Vol. 16, No. 4, December 1965, p. 411.

8.11. Standards for Quality Programs in Agricultural/Agribusiness Education, Part F, Section 553 Grant, Agricultural Education Department, Iowa State University, Ames, 1977.

8.12. Oren, J. W., et al., An Analysis of Standards for Teacher Education Programs in Agriculture/Agribusiness, report of the AATEA ad hoc committee on the Legitimization of Program Standards and Guidelines, published by Mississippi State University, Mississippi State, 1977.

8.13. An Assessment of the Pedagogical Skills Taught to Agricultural Education Undergraduates, report of the American Association of Teacher Educators in Agriculture ad hoc committee on Teaching Techniques, 1977.

8.14. Achieving the Potential for Performance-based Teacher Education: Recommendations, PBTE Series No. 16, American Association of Colleges for Teacher Education, Washington, D.C., 1974, pp. 32-33.

8.15. Broudy, H. S., and Applegate, W. K. "A Cautionary Appraisal of CBTE," The Journal of the American Association of Teacher Educators in Agriculture, Vol. 16, No. 3, November 1975.

8.16. Hillison, J. "CBTE, Like Wine . . . Best in Moderation," The Journal of the American Association of Teacher Educators in Agriculture, Vol. 20, No. 2, July 1979, pp. 22-26.

8.17. Broudy, H. D. Paper presented at Region V Conference on Competency-based Teacher Education, Chicago, March 1975.

8.18. Peterson, R. L. PBTE in Agricultural Education: A State of the Art, American Association of Teacher Educators in Agriculture Committee on Teacher Education in Agriculture Guidelines, 1975.

Chapter 9

9.1. Conant, J. B. The Education of American Teachers, New York: McGraw-Hill Book Company, 1963, p. 210.

9.2. Cogan, M. L. Clinical Supervision, Boston: Houghton Mifflin Company, 1973.

9.3. Teacher Education: The Decade Ahead, report of the Dekalb Conference on Teacher Education, National Education Association, Washington, D.C., 1955, p. 114.

9.4. Standards for Quality Programs in Agricultural/Agribusiness Education, Agricultural Education Department, Iowa State University, Ames, 1977.

9.5. *Summer Practice for Prospective Teachers of Vocational Agriculture*, Agricultural Education Department, Virginia Polytechnic Institute and State University, Blacksburg, 1966, pp. 2-3.

9.6. Wehrer, R. A. *A Systematized Field Experience Program: Pre-service in Secondary Education*, Gannan College, Erie, Pennsylvania, 1976.

9.7. Hemp, P. E. "Preparing the Teacher of Tomorrow," *The Agricultural Education Magazine*, Vol. 49, No. 4, October 1976, p. 89.

9.8. O'Kelley, G. L., Jr. *Agricultural Education: The Preparation of Teachers*, Vocational Education Bulletin No. 295, Agricultural Series No. 78, U.S. Office of Education, Washington, D.C., U.S. Government Printing Office, 1962, pp. 6-7.

9.9. Hemp. *Op. cit.*, p. 89.

9.10. Anderson, G. *Guidelines for Calling Student Teaching Case Conferences*, mimeograph, 1979.

9.11. Peterson, R. L. *A Handbook for Student Teachers and Supervising Teachers in Vocational Agriculture*, University of Minnesota, St. Paul, 1978.

9.12. Zubrick, O. J., Mortenson, J., and Peterson, R. *The Legimitization of Program Standards and Guidelines*, 1977.

9.13. Luft, V. D. *A Handbook for Student Teachers and Supervising Teachers in Vocational Agriculture*, North Dakota State University, Fargo, 1976.

9.14. Schuman, H. "Better University Supervisors and Cooperating Teachers," *The Agricultural Education Magazine*, Vol. 49, No. 4, October 1976, p. 81.

9.15. Stufflebeam, D. L. *Educational Evaluation and Decision Making in Education*, Itasca, Illinois: F. E. Peacock Publishers, Inc., 1971.

9.16. Beamer, R. W. "The Curriculum: Student Experience Programs," in V. R. Cardozier, ed., *Teacher Education in Agriculture*, 1st ed., Danville, Illinois: The Interstate Printers & Publishers, Inc., 1967, p. 231.

9.17. *The Purdue Agricultural Education Newsletter*, Vol. 1, Purdue University, West Lafayette, Indiana, 1978-79.

9.18. Hemp. *Op. cit.*, p. 90.

Chapter 10

10.1. Brown, R. D. *Student Development in Tomorrow's Higher Education: A Return to the Academy*, Student Personnel Series No. 16, American College Personnel Association, Washington, D.C., 1972, p. 28.

10.2. Williamson, E. G. The Student Personnel Point of View, Series VI, No. 13, American Council on Education, Washington, D.C., 1949, pp. 7-8.

10.3. Mayhew, L. B. Quoted in Chronicle of Higher Education, July 22, 1968, pp. 2, 4.

10.4. Sanford, N. Where Colleges Fail, San Francisco: Jossey-Bass, Inc., Publishers, 1967, p. 9.

10.5. Wrenn, C. G. Student Personnel Work in College, New York; Ronald Press, 1951, p. 27.

10.6. Cowley, W. H. "Reflections of a Troublesome but Hopeful Rip Van Winkle," Journal of College Student Personnel, Vol. 6, 1964, p. 68.

10.7. Brumbaugh, A. J., and Berdie, R. F. Student Personnel Programs in Transition, Series VI, No. 16, American Council on Education, Washington, D.C., October 1952, pp. 8-10.

10.8. Williamson. Op. cit., pp. 11-13.

10.9. Alpha Tau Alpha Official Manual, 3rd ed., Alpha Tau Alpha, The Pennsylvania State University, University Park, 1978, p. 3.

10.10. Ibid., pp. 4-7.

10.11. Teacher Education Groups in Agricultural Education Directory, National FFA Center, Alexandria, Virginia, undated. (Received on March 1, 1979)

10.12. 1981 Official FFA Manual for the National Organization for Students of Vocational Agriculture/Agribusiness, National FFA Center, Alexandria, Virginia, 1978, p. 114.

10.13. Ibid.

10.14. Teacher Education Groups . . ., op. cit.

10.15. Wolkow, B. S. Personal communication, March 2, 1979.

10.16. Governance Documents of Student National Education Association, Student National Education Association, Washington, D.C., 1978, p. 2.

10.17. Lee, J. S. Personal communication, February 13, 1979.

10.18. Stenzel, S. Personal communication, February 16, 1979.

10.19. Leroy, W. E. Personal communication, November 7, 1979.

Chapter 11

11.1. Good, C. V. Dictionary of Education, 3rd ed., New York: McGraw-Hill Book Company, 1973, p. 294.

11.2. Pennsylvania State Plan for Vocational Education, Pennsylvania State Department of Education, Harrisburg, 1979.

11.3. New York In-service Needs Assessment, Agricultural Education Division, Department of Education, Cornell University, Ithaca, New York, 1980.

11.4. True, A. C. A History of Agricultural Education: 1785-1925, Miscellaneous Publication No. 36, USDA, Washington, D.C., U.S. Government Printing Office, 1929, p. 273.

11.5. Martin, W. H. "Development of Teacher Education in Agriculture," in V. R. Cardozier, ed., Teacher Education in Agriculture, 1st ed., Danville, Illinois: The Interstate Printers & Publishers, Inc., 1967, p. 6.

11.6. True. Op. cit., p. 353.

11.7. Udell, G. G., comp. Laws Relating To Vocational Education and Agricultural Extension Work, Section 703, Public Law 347, 64th U.S. Congress, Washington, D.C., U.S. Government Printing Office, pp. 4-11.

11.8. Williams, J. F., Jr. Vocational Agriculture in Florida, Vol. 40, No. 2, Florida Department of Agriculture, Tallahassee, 1930, p. 292.

11.9. Shinn, G. C., Harris, E., Pruitt, A., and Witt, H. Inservice Teacher Education in Agricultural Education, report of an ad hoc committee of the American Association of Teacher Educators in Agriculture, 1979, pp. 3-4.

11.10. Ibid., p. 5.

11.11. Williams. Op. cit., p. 292.

11.12. Moore, G. "Innovations in Education," presentation made to the AATEA general meeting, Dallas, Texas, December 3, 1978.

11.13. Good. Op. cit., p. 126.

11.14. Ibid., p. 652.

11.15. Ibid., p. 151.

11.16. Craig, R. L., ed. Training and Development Handbook: A Guide to Human Resource Development, 2nd ed., New York: McGraw-Hill Book Company, 1976.

11.17. Cushman, H. R. Guidelines for In-service Education, unpublished UNESCO paper, FAO, Rome, 1980, pp. 2-3.

11.18. Shinn. Op. cit., p. 2.

Chapter 12

12.1. Schumann, H. "Individual Responsibility for Professional Input," The Agricultural Education Magazine, Vol. 48, May 1976, pp. 243, 244.

12.2. Wasserman, E. R., and Switzer, E. E. The Random House Guide to Graduate Study, New York: Random House, Inc., 1967, p. 3.

12.3. Warmbrod, J. R. "A Goal for the Next Decade: Quality Programs in Agricultural Education," The Agricultural Education Magazine, Vol. 50, June 1978, pp. 268-270.

12.4. Attaway, J. P. "Why I Planned a Graduate Program in Agricultural Education," The Agricultural Education Magazine, Vol. 35, May 1963, p. 231.

12.5. Bail, J. P. "Planning for the Master's Degree," The Agricultural Education Magazine, Vol. 35, May 1963, p. 232.

12.6. Bjoraker, W. T. "Graduate Programs Must Be Individualized," The Agricultural Education Magazine, Vol. 35, May 1963, p. 229.

12.7. Broyles, W. A. Graduate Work in Agricultural Education, Ph.D. thesis, University of Illinois, Urbana, 1925; published in The Pennsylvania State College Bulletin, State College, June 1926, p. 35.

12.8. Grigg, C. M. "Who Wants to Go to Graduate School and Why," Research Reports in Social Science, Vol. 11, No. 1, the Florida State University Center for Social Research, Tallahassee, February 1959.

12.9. LaRue, L. "Why Don't Vo-Ag Teachers Get a Master's Degree?," The Agricultural Education Magazine, Vol. 36, December 1963, pp. 160-161.

12.10. Allen, L. D. "Professionalism--Who Needs It? We All Do!" The Agricultural Education Magazine, Vol. 51, December 1978, pp. 123, 135.

12.11. DeBertin, R. "Professionalism and the Agriculture Teacher," The Agricultural Education Magazine, Vol. 51, December 1978, p. 130.

12.12. Broyles. Op. cit., pp. 37-41.

12.13. Wintebourne, R. J. "Full Year Internship vs Fifth Year Program," The Agricultural Education Magazine, Vol. 47, August 1974, pp 29, 40.

12.14. Eells, W. C. Degrees in Higher Education, New York: Center for Applied Research in Education, Inc., 1963, p. 106.

12.15. Carter, R. J. "Professional Competencies Needed and Possessed by Beginning Teacher Educators in Agricultural Education," unpublished Ph.D. thesis, Iowa State University, Ames, 1975.

12.16. Binkley, H. R. "Teacher Education Programs in Agricultural Education Should Be Located in Colleges of Education," The Journal of the American Association of Teacher Educators in Agriculture, Vol. 18, November 1977, pp. 2, 4-6, 26.

12.17. Knebel, E. H. "Teacher Education Programs in Agricultural Education Should Be Located in Colleges of Agriculture," _The Journal of the American Association of Teacher Educators in Agriculture_, Vol. 18, November 1977, pp. 3, 7-10, 32.

12.18. Simpson, R. H. "Adapting Instruction to Individual Differences," _The Agricultural Education Magazine_, Vol. 32, August 1959, pp. 40-43.

12.19. _Ibid._

12.20. Hamlin, H. M. "The Teacher of Agriculture as Thinker and Scholar," _The Agricultural Education Magazine_, Vol. 30, September 1957, p. 60.

12.21. Johns Hopkins University Register, 1880-81, p. 21, quoted in W. C. Ryan, _Studies in Early Graduate Education_, New York: Arno Press, Inc., and _The New York Times_, 1971.

12.22. Eells. _Op. cit._, p. 106.

12.23. Kent, L. J., and Springer, G. P., eds. _Graduate Education Today and Tomorrow_, Albuquerque: University of New Mexico Press, 1972, p. 104.

12.24. Jordan, T. E. "Innovation and the Master's Degree," proceedings of the thirty-second annual meeting of the Midwestern Association of Graduate Schools, Raymond P. Maricella Edition, Loyola University, Chicago, 1976, p. 6.

12.25. Krebs, A. H. "Graduate Study for Teachers of Agriculture," in V. R. Cardozier, ed., _Teacher Education in Agriculture_, 1st ed., Danville, Illinois: The Interstate Printers & Publishers, Inc., 1967, p. 316.

Chapter 13

13.1. _Guiding Principles for Institutions Training Teachers of Vocational Agriculture_, report of the AATEA Teacher Education Committee at the annual conference of the American Vocational Association, Buffalo, New York, August 1958.

13.2. Cross, A. _Home Economics Evaluation_, Columbus, Ohio: Charles E. Merrill Publishing Company, 1973.

13.3. Evans, R. N., and Terry, D. R. _Changing the Role of Vocational Teacher Education_, Bloomington, Illinois: McKnight Publishing Company, 1971, p. 184.

13.4. Anderson, S. B., _et al._ "Evaluation Concepts," _Encyclopedia of Educational Evaluation_, San Francisco: Jossey-Bass, Inc., Publishers, 1975.

13.5. _Evaluation as Feedback and Guide_, report of the ASCD 1967 Yearbook Committee, Association for Supervision and Curriculum Development, National Education Association, Washington, D.C., 1967, pp. 4-7.

13.6. Tuckman, B. W. *Evaluating Instructional Programs*, Boston: Allyn & Bacon, Inc., 1979, pp. 10-11, 199-200.

13.7. Knuti, L. L. *Leadership Education*, Bozeman: Montana State University, 1965.

13.8. Tuckman. *Op. cit.*, p. 12.

13.9. Coster, J. K. "Program Evaluation," *Review of Educational Research*, Vol. 38, 1968, pp. 417-433.

13.10. *Standards for Quality Programs in Agricultural/Agribusiness Education*, Agricultural Education Department, Iowa State University, Ames, 1977.

13.11. *Accredited Postsecondary Institutions and Programs*, U.S. Office of Education, Washington, D.C., U.S. Government Printing Office, 1973.

13.12. *Nationally Recognized Accreditating Agencies and Associations*, U.S. Office of Education, Washington, D.C., U.S. Government Printing Office, 1977.

13.13. *Standards for the Accreditation of Teacher Education*, National Council for Accreditation of Teacher Education, Washington, D.C., 1979.

13.14. Haberman, M., and Stinnett, T. M. *Teacher Education and the New Profession of Teaching*, Berkley, California: McCutchan Publishing Corporation, 1973, p. 249.

13.15. *Standards for State Approval of Teacher Education*, National Association of State Directors of Teacher Education and Certification, Salt Lake City, 1976, p. 37.

13.16. *Ibid.*, pp. 39-40.

13.17. *Guidelines for Developing Programs in Agricultural Education for the 1970's*, unpublished report, National Outlook Conference in Agricultural Education, 1969.

13.18. *Transitions in Agricultural Education Focusing on Agribusiness and Natural Resources Occupations*, American Vocational Association, Washington, D.C., 1976.

13.19. *Teacher Education in Agriculture Guidelines*, American Association of Teacher Educators in Agriculture, 1976.

13.20. *Standards for Quality Programs* . . ., *op. cit.*

13.21. *Guidelines for Developing Programs* . . ., *op. cit.*

13.22. *Transitions in Agricultural Education* . . ., *op. cit.*

13.23. *Teacher Education in Agriculture Guidelines*, *op. cit.*

13.24. *Standards for Quality Vocational Programs* . . ., *op. cit.*

13.25. *Ibid.*, pp. x-1 to x-25.

13.26. Tom, F. K. T., and Cushman, H. R. "The Cornell Diagnostic Observation and Reporting System for Student Description of College Teaching," Search, Agricultural Experiment Station, Cornell University, Ithaca, New York, Vol. 5, No. 8, November 1975, pp. 1-27.

13.27. Zurbrick, P. A Department Follows Up Its Graduates, 1967-1974, Department of Agricultural Education, The University of Arizona, Tucson, 1975.

13.28. Miller, L. E. "A Five-Year Follow-Up of Agricultural Education Graduates with Implications for Improving Recruitment and Retention Efforts," The Journal of the American Association of Teacher Educators in Agriculture, Vol. 14, No. 3, November 1973, pp. 23-25.

13.29. Boucher, L. W. "The Status of Teacher Education Programs in Agriculture," The Journal of the American Association of Teacher Educators in Agriculture, Vol. 13, No. 1, 1972, pp. 1-5.

13.30. Elam, S. Performance-based Teacher Education: What Is the State of the Art?, PBTE Monograph Series No. 1, American Association of Colleges for Teacher Education, Washington, D.C., 1971, p. 22.

13.31. Elam, S. A Résumé of Performance-based Teacher Education: What Is the State of the Art?, PBTE Monograph Series No. 1-a, American Association of Colleges for Teacher Education, Washington, D.C., 1972, p. 4.

Chapter 14

14.1. Stewart, B. R., Richardson, W. B., and Shinn, G. C. "Problems of the Profession Needing Attention," Research Conference in Agricultural Education Proceedings: Central Region, The Ohio State University, Columbus, August 1976, pp. 129-148.

14.2. Mannebach, A. J. "An Analysis of the Impact of Agricultural Education Research on Identified Professional Concerns," The Journal of the American Association of Teacher Educators in Agriculture, Vol. 21, No. 1, pp. 19-25.

14.3. Drake, W. E. Developing Programmatic Research in Agricultural Education, paper presented at the annual Southern Research Conference in Agricultural Education, Starkville, Mississippi, July 1979.

14.4. Krebs, A. H. "Research in Agricultural Education from a Different Perspective," Research Conference in Agricultural Education Proceedings: Central Region, The Ohio State University, Columbus, August 1976, pp. 3-10.

14.5. Warmbrod, J. R., and Miller, L. E. Research Activities in Teacher Education, unpublished manuscript, Department of Agricultural Education, The Ohio State University, Columbus, 1979.

14.6. Patterson, S. D. "Research in Agricultural Education Should Be Handled by Units Separate from Teacher Education Departments," *The Journal of the American Association of Teacher Educators in Agriculture*, Vol. 20, No. 3, 1979, pp. 3-7; 52-53.

14.7. Bowen, B. E. *Job Satisfaction of Teacher Educators in Agriculture*, paper presented at the National Agricultural Education Research meeting, New Orleans, December 1980.

14.8. Osborne, E. *Requirements of Masters' Degree Candidates in the Non-thesis Option* (Independent Study), unpublished manuscript, Department of Agricultural Education, The Ohio State University, Columbus, December 1980.

14.9. Mannebach. *Op. cit.*

14.10. Krebs. *Op. cit.*

14.11. Newcomb, L. H. *Agricultural Education: Review and Synthesis of the Research*, Information Series No. 139, National Center for Research in Vocational Education, Columbus, Ohio, 1978.

14.12. Warmbrod, J. R., and Phipps, L. J. *Review and Synthesis of Research in Agricultural Education*, Center for Research and Leadership Development in Vocational and Technical Education, Columbus, Ohio, August 1966.

14.13. Carpenter, E. T., and Rodgers, J. H. *Review and Synthesis of Research in Agricultural Education*, Center for Vocational and Technical Education, Columbus, Ohio, June 1970.

14.14. Warmbrod and Miller. *Op. cit.*

14.15. Arthur, P., and Budke, W. E. *Current Projects in Vocational Education, FY 1978*, National Center for Research in Vocational Education, Columbus, Ohio, January 1980.

14.16. Arthur, P., and Budke, W. E. *Current Projects in Vocational Education, FY 1979*, National Center for Research in Vocational Education, Columbus, Ohio, June 1980.

14.17. Warmbrod and Miller. *Op. cit.*

14.18. Krebs. *Op. cit.*

14.19. Taylor, R. E. "Improving Research in Departments of Agricultural Education," *Research Conference in Agricultural Education Proceedings: Central Region*, The Ohio State University, Columbus, August 1976, pp. 84-95.

14.20. Brown, R. A. "Research in Agricultural Education Should Be Handled by Teacher Education Departments," *The Journal of the American Association of Teacher Educators in Agriculture*, Vol. 20, No. 3, 1979, pp. 2; 8-10.

14.21. Copa, G. H. What Should We Do About Research in Vocational Education?, paper presented at the annual meeting of the American Vocational Education Research Association, New Orleans, December 1980.

14.22. Ibid.

14.23. Brown. Op. cit.

14.24. Oliver, J. D. Improving Agricultural Education Research, paper presented at the National Agricultural Education Research meeting, Anaheim, California, November 1979.

14.25. Oliver, J. D., and Hinkle, D. E. Selecting Inferential Statistics, paper presented at the Southern Research Conference in Agricultural Education, Auburn, Alabama, July 1980.

14.26. Brown. Op. cit.

14.27. Warmbrod, J. R. "Reporting Research at Conferences and Conventions," The Journal of the American Association of Teacher Educators in Agriculture, Vol. 14, No. 3, 1973, pp. 12-17.

14.28. Copa, G. H. "Key Questions in Evaluating and Reporting Research in Vocational Education," The Journal of Vocational Education Research, Vol. 3, No. 2, 1978, pp. 1-6.

14.29. Miller, L. E. "Preparing a Research Journal Article," The Journal of the American Association of Teacher Educators in Agriculture, Vol. 20, No. 1, 1979, pp. 32-36.

14.30. Newcomb. Op. cit.

14.31 Krebs. Op. cit.

14.32. Bowen. Op. cit.

14.33. Newcomb. Op. cit.

14.34. Copa. What Should We Do About Research in Vocational Education?, op. cit.

14.35. Brown. Op. cit.

Chapter 15

15.1. Thuemmel, W. L., and Welton, R. F. AATEA Survey of Teacher Education Activity in International Agriculture, unpublished results of a national study, February 1979.

15.2. Bender, R. E. "The Program of Agricultural Education with Implications for Colleges of Agriculture and Natural Resources," in D. L. Armstrong, ed., Impact of Enrollments and Student Body Composition on Academic Programs, Design, and Delivery, RICOP report, Michigan State University, East Lansing, 1977, p. 430.

15.3. Agricultural Education for the Seventies and Beyond, report prepared by the Committee on Agricultural Education, Commission on Education in Agricultural and Natural Resources, National Research Council, and published by the American Vocational Association, Washington, D.C., July 1971, pp. 30-31.

15.4. Ibid.

15.5. Turk, K. L., Snyder, D. E., and Scott, J. T. Rationale for Involvement of Universities from the United States of America in International Education, Research, and Development, Publication No. 2, Association of United States University Directors of International Agricultural Programs, March 1979, p. 3.

15.6. Hannah, H. W. Resource Book for Rural Universities in the Developing Countries, Urbana: University of Illinois Press, 1966, p. 12.

15.7. Matteson, H. R. "Involvement in International Agriculture--A Challenge for Agricultural Education," The Agricultural Education Magazine, Vol. 50, 1978, p. 195.

15.8. Thuemmel, W. L. "Beyond Filling the World's Breadbasket: A Challenge for Agricultural Education," The Agricultural Education Magazine, Vol. 52, 1979, pp. 3-4; 7.

15.9. Bergland, B. "Global Interdependence in Meeting World Food Needs," remarks prepared by the U.S. Secretary of Agriculture for delivery at the twentieth FAO Conference, Rome, November 13, 1979, p. 8.

15.10. Smith, M. G. "International Agriculture Training," in D. L. Armstrong, ed., Impact of Enrollments and Student Body Composition on Academic Programs, Design, and Delivery, RICOP report, Michigan State University, East Lansing, 1977, p. 300.

15.11. Wortman, S., and Cummings, R. W., Jr. To Feed This World: The Challenge and the Strategy, Baltimore: The Johns Hopkins University Press, p. 397.

15.12. Ibid., p. 400.

15.13. Smith. Op. cit., p. 305.

15.14. Trotter, E. E. Directory of Agricultural Teacher Educators, compiled in cooperation with the American Association of Teacher Educators in Agriculture, 1979-80.

15.15. Rawlings, N. O., and Foutch, H. W. "NACTA's Interest and Experience," National Association of Colleges and Teachers of Agriculture Journal, Vol. 21, No. 2, 1977, pp. 20-22.

15.16. McCreight, D. E. "Assessment of Your Potential for International Employment," The Agricultural Education Magazine, Vol. 52, 1979, pp. 134-135.

Chapter 17

17.1. Vocational Education Act of 1963, Public Law 88-210, 88th U.S. Congress, Washington, D.C., U.S. Government Printing Office, 1963.

17.2. President's Panel of Consultants on Vocational Education. Education for a Changing World of Work, O.E. 800-21, U.S. Office of Education, Washington, D.C., U.S. Government Printing Office, 1963.

17.3. Smith-Hughes Act, Publication No. 347, 64th U.S. Congress, Washington, D.C., U.S. Government Printing Office, 1917.

17.4. Hunsicker, N. H. "New Challenges in Teacher Education and Supervision," unpublished speech, 1967.

17.5. Smith-Hughes Act, op. cit.

The Authors

MAX L. AMBERSON is professor in and head of the Department of Agricultural Education at Montana State University. He has taught vocational agriculture in Montana and has been state supervisor of agricultural education, state director of vocational education, and assistant director of the Peavey Grain Company. He received his B.S. from Montana State University, his M.A. from the University of Minnesota, and his Ph.D. from The Ohio State University.

WILLIAM H. ANNIS is professor in and chairperson of the Occupational Education Department at the University of New Hampshire. He is a former supervisor of teacher training with the Massachusetts State Department of Education. He received his Ed.D. from Cornell University in 1961.

JOE P. BAIL is chairperson of the Department of Education at Cornell University, where he holds a professorship in agricultural education. His current position involves the administration of an education department with 40 professional staff persons. He has been a teacher educator at West Virginia University, Michigan State University (Ph.D. 1958), and Cornell University. Inservice education and instructional materials development have always been a major component of his professional work.

ARTHUR L. BERKEY is professor of agricultural and occupational education at Cornell University. He taught vocational agriculture and was a high school principal in Michigan. He received his Ph.D. from Michigan State University in 1967.

DOUGLAS BISHOP is professor of agriculture at Montana State University and acting associate dean of the college of agriculture. He taught vocational agriculture in Colorado for a number of years. He worked for two years as a research associate at the National Center for Research in Vocational Education while he was completing his Ph.D. at The Ohio State University. He is co-author of two texts on crop science.

RONALD A. BROWN is associate professor of agricultural and extension education at Mississippi State University. He also serves as editor of The Journal of the American Association of Teacher Educators in Agriculture. He received his Ph.D. at the University of Illinois in vocational and technical education, with an emphasis on agricultural education. He has authored and co-authored a number of books and articles.

CHARLES W. BYERS is professor of agricultural education in the Department of Vocational Education at the University of Kentucky. He has received numerous awards for his excellence in teaching and advising. He is a recognized leader in student organizations in vocational education. His Ph.D. was earned at The Ohio State University.

JAMES P. CLOUSE is professor of agricultural education at Virginia Polytechnic Institute and State University. He served as secretary of the Agricultural Education Division of the American Vocational Association for three years and as vice president of the American Association of Teacher Educators in Agriculture for another three years. He received his Ph.D. from Purdue University. He has written extensively on the guidance role of the vocational teacher and is very interested in international agricultural education programs.

JOHN R. CRUNKILTON is professor and program area leader of agricultural education at Virginia Polytechnic Institute and State University. He has served as editor of _The Journal of the American Association of Teacher Educators in Agriculture_ and as president of the AATEA. He received his Ph.D. from Cornell University. He has co-authored several books in the areas of human relations, curriculum development, and problem solving.

CHARLES C. DRAWBAUGH is professor in and past chairperson of the Department of Vocational-Technical Education in the Graduate School of Education at Rutgers University. He has been both secretary and president of the American Association of Teacher Educators in Agriculture. He earned his Ed.D. degree at The Pennsylvania State University. He has co-authored the book, _Agricultural Education: Approaches to Teaching and Learning_.

PAUL E. HEMP is professor in and chairperson of the Division of Agricultural Education at the University of Illinois. He has served as a teacher educator at the University of Vermont and at Purdue University and has been on the staff of the University of Illinois since 1958.

ALFRED J. MANNEBACH is professor of agricultural education at the University of Connecticut. His past activities include eastern regional vice president of the American Association of Teacher Educators in Agriculture and editor of the 1976-77 and 1977-78 _Summaries of Research and Development Activities in Agricultural Education_. He received his Ed.D. degree from the University of Illinois. He is recognized for his work in educational program evaluation.

W. HOWARD MARTIN is professor emeritus of agricultural education at the University of Connecticut. A former editor of _The Agricultural Education Magazine_, he has long been a prominent leader in agricultural education. He earned his Ed.D. at the University of Illinois, where his dissertation dealt with interpretations of the Smith-Hughes Act.

R. PAUL MARVIN is professor in and head of the Agricultural Education Division at the University of Minnesota. The World Bank and the International Labor Organization have utilized his services as a consultant in vocational education for rural development. He received his Ph.D. from the University of Minnesota, with a major in education and a minor in educational administration.

FLOYD G. McCORMICK is professor in and head of the Department of Agricultural Education at The University of Arizona. He is a past president of the American Association of Teacher Educators in Agriculture. His Ph.D. was awarded by The Ohio State University. He is an advocate of quality programs which are based upon a sound philosophical foundation.

J. DAVID McCRACKEN is professor of agricultural education at The Ohio State University. Prior to assuming this position, he was a research and development specialist and assistant director at the National Center for Research in Vocational Education. He has served as president of the American Association of Teacher Educators in Agriculture and as editor of the *Journal of Vocational Education Research.* He received his bachelor's and master's degrees from Iowa State University and his Ph.D. from The Ohio State University.

DONALD E. MCCREIGHT is professor of agricultural and extension education at the University of Rhode Island. His past activities include being a Peace Corps instructor, a USAID resident project leader in the Azores, and a guest lecturer for international extension seminars in the Netherlands and in West Germany. Currently, his teaching duties include a graduate course in international fisheries and one in agricultural extension programs.

LARRY E. MILLER is professor in the Department of Agricultural Education at The Ohio State University and is currently involved in teaching research methodology and design. He was formerly on the faculties at the University of Missouri - Columbia and at the Virginia Polytechnic Institute and State University. He has also served as editor of *The Journal of the American Association of Teacher Educators in Agriculture.* He holds a Ph.D. from Purdue University.

CLIFFORD L. NELSON is professor in and chairperson of the Department of Agricultural and Extension Education at the University of Maryland. His previous position was on the faculty at the University of Minnesota, where he completed his Ph.D. His past services include being a resident consultant in Brazil, a former secretary and vice president of the American Association of Teacher Educators in Agriculture, a former president of the board of *The Agricultural Education Magazine*, an associate editor of *The Journal of the American Association of Teacher Educators in Agriculture*, and an agricultural representative to the *AVA Journal* editorial board. He is currently the president-elect of the American Association of Teacher Educators in Agriculture.

NICHOLAS L. PAUL is chief of the Division of Marketing in the North Carolina State Department of Agriculture. Before assuming that position, he was chairperson of the Occupational Education Department at the University of New Hampshire. He is also president of Interstate Associated, Inc., an educational consulting firm in Raleigh, North Carolina. He received his Ed.D. from North Carolina State University in 1973.

EDGAR A. PERSONS is professor of agricultural education at the University of Minnesota. His primary work responsibilities are in the area of adult education, with special emphasis on management education programs for farmers, loggers, and small business owners. He received his Ph.D. from the University of Minnesota in 1966, with a major in education and a minor in educational psychology.

ROLAND L. PETERSON is associate professor of agricultural education at the University of Minnesota. Formerly, he was on the faculty at The University of Nebraska. He has also been an assistant supervisor for the Nebraska State Department of Education. He has developed a competency-based undergraduate teaching methods course for agricultural education majors. He received his Ed.D. from The University of Nebraska.

C. CAYCE SCARBOROUGH is professor emeritus at North Carolina State University and at Auburn University. He was head of agricultural education at North Carolina State University from 1949 to 1973 and head of agricultural education at Auburn University from 1973 to 1979. He also served as editor of <u>The Agricultural Education Magazine</u> from 1965 to 1970. He received his Ed.D. from the University of Illinois in 1950.

GLEN C. SHINN is professor of agricultural and extension education and agricultural engineering at Mississippi State University. He has been coordinator of the inservice education program there, besides being a specialist in agricultural mechanics. He has also been the chairperson of the AATEA ad hoc committee on inservice education. He earned his Ph.D. at the University of Missouri, where his doctoral thesis dealt with the relationship of classroom and laboratory time and the performance of adults.

GORDON I. SWANSON is professor of agricultural education and director of graduate studies in the Department of Vocational Education at the University of Minnesota. He is a past president of the American Vocational Association, Phi Delta Kappa International, and Section Q (Education) of the American Association for the Advancement of Science. An active researcher and contributor to the literature of the field, he received the Distinguished Service Award of the National Center for Research in Vocational Education at The Ohio State University in 1977. He received his Ph.D. from the University of Minnesota.

JOHN F. THOMPSON is professor in and chairperson of the Department of Continuing and Vocational Education at The University of Wisconsin - Madison. He earned his Ph.D. at Michigan State University. He is the author of <u>Foundations of Vocational Education: Social and Philosophical Concepts</u>.

WILLIAM L. THUEMMEL is associate professor and head of agricultural education at the University of Massachusetts. He has conducted agricultural education research in Michigan and Massachussets, as well as overseas in Guam, Jamaica, and Taiwan. His dissertational research dealt with vocational agriculture school graduates in Taiwan. From 1970 to 1974, he served as a teacher educator and land-grant program implementation specialist at the University of Guam. He received his Ph.D. from Michigan State University.

J. ROBERT WARMBROD is professor in and chairperson of the Department of Agricultural Education at The Ohio State University. He is a past president of the American Vocational Education Research Association and a vice president of the American Vocational Association for Agricultural Education. He earned his Ed.D. at the University of Illinois, where he was a faculty member prior to going to The Ohio State University in 1968.

RICHARD F. WELTON is associate professor of agricultural education at Kansas State University. He is responsible for coordinating the preservice program. His international experiences include working with the United Nations Development Program project in Brazil from 1971 to 1973 and working as a consultant to Yemen in another United Nations project in 1977. His Ph.D. degree is from The Ohio State University.

Index

F

G

H

I

R

S

T

U

V

W